From Bogotá to Beijing

From Bogotá to Beijing

Development and Life after Globalization

David Jacoby

LEXINGTON BOOKS
Lanham • Boulder • New York • London

Published by Lexington Books
An imprint of The Rowman & Littlefield Publishing Group, Inc.
4501 Forbes Boulevard, Suite 200, Lanham, Maryland 20706
www.rowman.com

Unit A, Whitacre Mews, 26-34 Stannary Street, London SE11 4AB

British Library Cataloguing in Publication Information Available

Library of Congress Cataloging-in-Publication Data Available
Library of Congress Control Number: 2018955260

ISBN 978-1-4985-5674-3 (cloth : alk. paper)
ISBN 978-1-4985-5675-0 (electronic)
ISBN 978-1-4985-8864-5 (pbk. : alk. paper)

∞™ The paper used in this publication meets the minimum requirements of American National Standard for Information Sciences Permanence of Paper for Printed Library Materials, ANSI/NISO Z39.48-1992.

Printed in the United States of America

To Jessica, Brent, Weston, and Camille

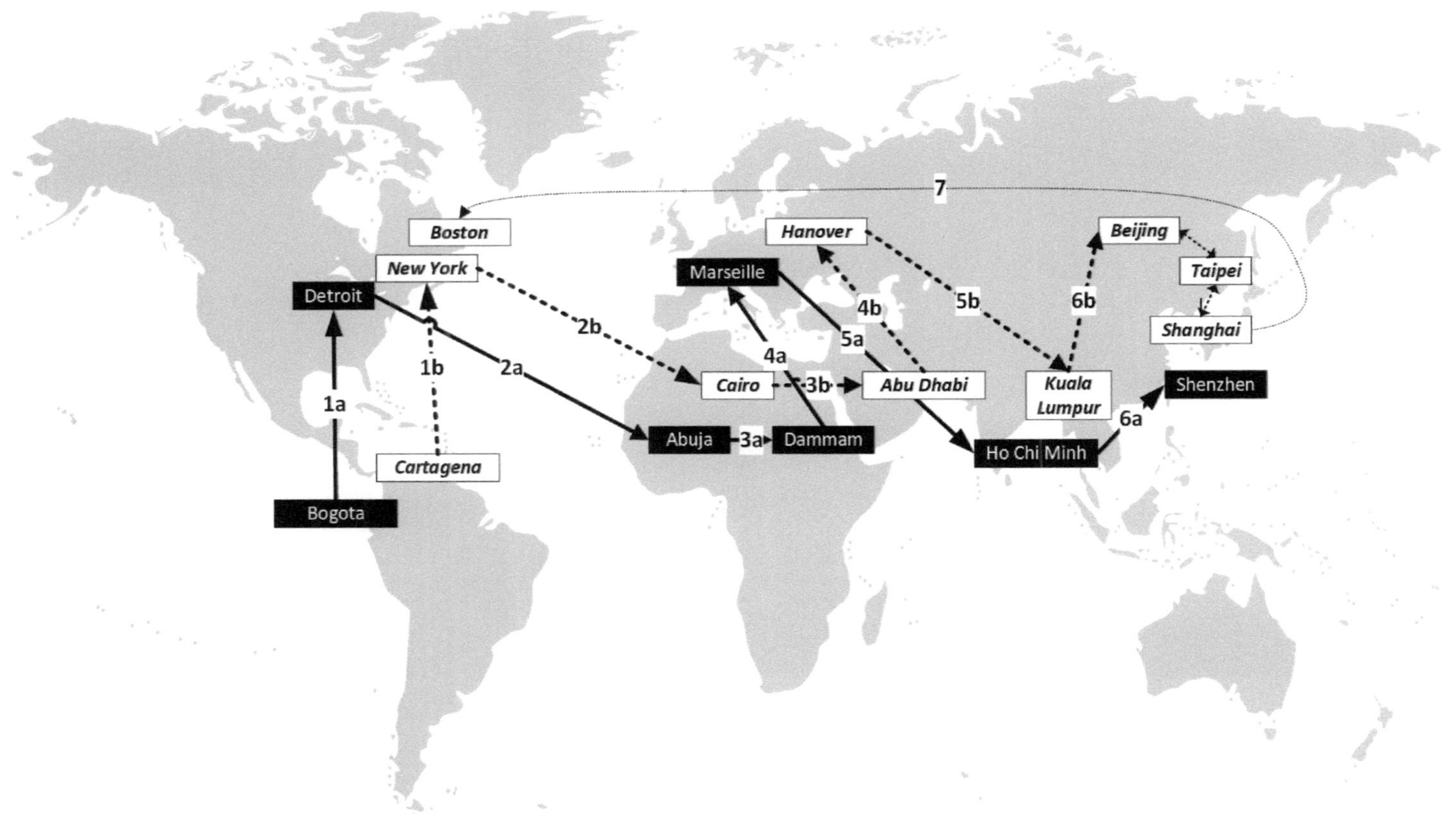

Figure 0.1. Map

Contents

Foreword

In our times, when practically every inhabitant of the world is aware that there are no more undiscovered lands in our planet, yet many new settings and challenges for the untiring imagination of human beings and for the ways in which we trade and relate to each other, in an unquestioned context of globalization, we must ask ourselves: What is next?

The extraordinary traveling odysseys and varied endeavors around the world relayed in this book supply us with sound guidance on how to start formulating the right answer or better still, the right and multidimensional answers in plural.

David Jacoby's genuine and keen curiosity, his description of almost countless venues and persons of all sorts of life in many places in the world are truly amusing and enlightening in both their common characteristics and their heterogeneous qualities of world cultures and specific human personalities.

He shows a young spirit, amazed at simple and complex events of life, while posing equally simple, but profound questions dealing with the future after globalization and the corresponding patterns of production and consumption that have shaped a new global state of things, and wisely concludes that "the growth we are enjoying is unsustainable and has come at a price."

He acknowledges that "nobody violated any rules, or even did anything wrong. There are no fingers to point." Yet, he adds, "we have collectively been somewhat ecologically, environmentally, and socially reckless, and we are now at a turning point."

In a way this "turning point" marks the end of globalization but Jacoby is "disquieted by the feeling that the world has quite a few challenges managing growth and development." In order to meet them, his book becomes a most powerful tool and the judicious assemblage of his set of personal accounts, insightful inferences, interpretations, tapestry of anecdotes, and roadmap to a better life is thoughtful, important, and timely. It can make a difference in the way whole cultures live, work, and play across national and cultural boundaries in the future.

This book uses the complex prism of international business to distill a lifetime of travel experiences into a solution to the biggest challenge of our time. It is an economist's equivalent of *The Great Gatsby*, weaving an important tale about the confluence of economic, financial, cultural, political, and social trends that have led to an era of excess, which has become intertwined with what we know as globalization. But Jacoby does not stop at capturing the problem articulately and artistically. He shows us the way to a better life, and he does it with granular specificity that compels us to action. The roadmap is clear, and the book is a must-read for anybody who is concerned about the impact of development and where we should go "after globalization."

Miguel G. Breceda Lapeyre
Mexico City, April 2018.

Preface

Business is a social science. To be sure, there are cut-and-dried functional disciplines, such as accounting, finance, and operations research, and even business strategy can be approached in a scientific way that optimizes portfolios and diversifies risks. Yet even in these disciplines, experienced experts in these areas know that the implementation of any business practice or theory depends on human motivation and understanding, personal and organizational goals, and vision for the future, and not least, alignment between many people. Without alignment, everybody can be working toward their own understanding, their own goals, and their own visions, with little or no collective progress, even regression.

Economics is similarly a social science. During the first two years of my economics studies, I learned to quantitatively balance supply with demand and investment with savings, to adjust money supply to achieve inflation targets, and to model deficits and national debt. However, in subsequent years, more and more of even the empirical, quantitative, studies dealt with behavior and perceptions. After all, the effectiveness of an economic policy is directly related to the extent to which people respond to a tax break, price change, or other stimulus. If companies and individuals pocket tax breaks instead of reinvesting the money, the outcome can be completely different than may have been intended. Consumers may be more inclined to buy an overpriced luxury good than an inexpensive one which offers a better value because of their perceptions and the image they wish to project to others.

International business, which involves both business *and* economics, in widely varying political, religious, and linguistics contexts, is the ultimate social science, and some call it an art. The practice and implementation of business strategies and economic policies varies widely across cultures. Habits and traditions affect the willingness to change, adapt, and respond—societies are slow to change, and the role of women in society and business greatly changes the scope, likelihood, and rate of business and economic change. The concentration of power and income in an organization or a society can totally change the relevance and effectiveness of an economic policy—for example, middle-class tax cuts can be meaningless, or nearly so, in an oligarchy. Differences in religious beliefs and aspirations can totally change whether people are even listening to, or care about, business or economics, profit or growth—different concepts of social obligation and afterlife make business and money central to some and anathema to others. Language differences can impede common understanding of even basic facts and concepts—for example, the words profit, value added, and partnership have very different connotations in various languages.

In a way, this book shares some of my experiences and reflections based on thirty years of international management consulting. It just begins to scratch the surface of the complex and multifaceted crystal through which different societies around the world perceive and engage in business and economics, and how this has changed due to globalization.

This book is a work of fiction. Any resemblance of the companies, products, or people in this book to real companies, products, brands, or people is purely coincidental. The value chain descriptions in this book are fictional. Neither the conversion steps described, not the conditions in specific countries, are guaranteed to be true to life or representative of the countries in which they are depicted. The D-Factor methodology is explained in my book *The High Cost of Low Prices: A Roadmap to Sustainable Prosperity* (Business Expert Press, 2017).

The stories are organized in two parallel and interwoven trips around the world. The first trip takes place in 2010, and goes from Bogotá to Detroit, to Lagos and Abuja, to Dammam, to Paris and Marseille, to Ho Chi Minh City, and ends in Shenzhen. The second trip takes place in 2030, and follows a similar trajectory. It starts in Cartagena and goes to New York, and on to Cairo, Abu Dhabi, Hanover, Kuala Lumpur, and Beijing (with side trips to Shanghai and Taipei), and ends in Boston. The two stories are interwoven, with parallel and similar legs of each trip alternating. The 2010 trip characterizes the way it was. The 2030 trip characterizes the way it can be. There is a theme that ties the first trip to the second trip; this is explored later in the book. An epilogue explains what happens later (in 2050) to the colorful characters that you meet on the trips.

An afterword summarizes some of the main themes and, in case you are interested, an appendix provides supplemental background on the European approach to balancing growth and progress through various social and environmental frameworks and policies.

If you have any feedback on the book, I would enjoy hearing from you. Please feel free to contact me at djacoby@bostonstrategies.com.

I hope you enjoy the story!

Acknowledgments

I would like to extend my gratitude to Joseph Parry and Emily Roderick at Lexington Press, who gave me the opportunity to take on this project.

I would also like to thank authors from multiple genres whose works have influenced the theme, structure, and style of this book. Thomas L. Friedman's critiques of today's society and perspectives on what to do about globalization (*Hot Flat, and Crowded* and so many other greats) set new benchmarks for substantive analysis and commentary. Paul Theroux's travel stories (*Dark Star Safari, The Old Patagonian Express, Riding the Iron Rooster*, and other greats) are about far more than travel; his work is unsurpassed in its deep content and rich interpretation of diverse cultural contexts, and it's fun to read. Annie Leonard's and Pietra Rivoli's creative and compelling documentaries (*The Story of Stuff* and *Travels of a T-Shirt*, respectively) have distilled the essence of complex and contemporary issues of development; their work has created awareness and encouraged action. Eli Goldratt's business novels like *The Goal* created an alternative genre that made deep analytical ideas and theories about production and supply chain management accessible to a broad audience.

Several economists have been inspirational to me over the years, and have shown me that economists can indeed make a positive impact on global development. Jeffrey Sachs, the renowned economist and benefactor, has been making a positive difference in the world since I graduated from college. Herman Daly, a champion of sustainability principles, provoked me into exploring unconventional understandings of growth, progress, and sustainability in an international world. Elinor Ostrom, who pioneered the area of collective action and won the Nobel prize in 2009, spent a lifetime trying to solve the right problem (she died in 2012). These three economists offer sometimes divergent approaches to economic development, and I learned from their differences.

While few politicians have inspired me over the years, the courage with which European leaders have identified, codified, and set standards for social, ecological, and environmental progress has earned my respect. They have successfully and responsibly used their positions to align broad and diverse segments of their constituents around collective action for the common good. They too have set a high bar for others to follow.

Leonard Lauder and the faculty of the Joseph H. Lauder Institute at the Wharton School have encouraged me to express my civic responsibility at the global level. I also appreciate the idealism, courage, and passion to improve the world displayed by my classmates at the institute, some of whose enthusiasm, energy, and idealism seems unbounded.

More fundamentally, my mother and father taught me right from wrong and to respect other people. My eighth grade teacher, Betts Louchheim, taught me to be an enthusiastic learner. Bob Waters, my ninth grade US History teacher, taught me civic responsibility by making me read and debate editorials in the *New York Times* every day.

Finally, I want to thank all the people in my business travels who have enriched my life over the years. I never expected an international consulting career to yield such a panoply of deeply respectful, affectionate—even loving, loyal, personally meaningful, and enduring relationships with which I have been blessed. This especially includes the young people such as Amit (India), Boon (Cambodia), and Phi (Vietnam), whom I hope will benefit from human-centered policies for development.

ONE

Bogotá to Detroit

BOGOTÁ, APRIL 24, 2010

Three of the overhead compartments flung open. Two backpacks and a duffel bag fell into the aisle. Our heads nearly banged into the seats in front of us. It seemed like a crash landing. But the captain didn't announce anything at all. The plane continued to move along the runway at an angle, then recovered. "Bienvenidos a Bogotá," announced the stewardess a minute later over the loudspeaker.

I was in Bogotá to give a presentation on pipeline security at an oil and gas symposium, and to present a consulting proposal to a major oil company, PetroFuerte. "Security" meant many things in the oil and gas business in Columbia. At a basic physical level, it meant structural integrity, usually preventing corrosion. At another level, it meant protecting the pipelines from thieves and terrorists.

The evening before the conference began, there was an exhibition. Milling around with a glass of wine, I ran into Gunther, who was manning his company's booth. He had some kind of display that demonstrated intelligent oilfield technologies on a number of computer screens. One of the screens was cracked and flickering. He explained apologetically to me that it was damaged during the robbery. Before leaving from Medellin for Cartagena, some thugs had secretly jumped into Gunther's car trunk, and they busted out when the car was in the mountains to steal their money and luggage and jump into a get-away car. After the robbery, the trunk wouldn't close so the equipment got wet.

The robbery wasn't Gunther's only problem. Gunther felt that he needed to prevent me from getting into similar troubles that he had recently experienced in Columbia, so he asked me if I planned on visiting "the other side of the river." Be very careful over there, he said. When you're done, he said, take the used condom with you. "Some of the girls impregnate themselves with your used condom and then come after you a year later demanding that you take financial responsibility for the child." I thanked Gunther for his advice and refreshed my drink.

At the conference the next morning, I delivered my pipeline security presentation. There were about seventy-five attendees. I gave my PowerPoint presentation and took questions, trying to sound like an expert on a topic that was not really in my core area of expertise. I had studied the topic for months before the conference, and I think I was sufficiently knowledgeable about it to achieve the objective. The moderator, who introduced me and facilitated the question and answer session, gave me a nice gift and shook my hand.

After the presentation and the chitchat with the six people that came up to the podium to talk to me about pipeline security, Juan Alejandro Muñoz Ramirez greeted me and gave me a bear hug. Though we hadn't seen each other often, Juan Alejandro was effusive and supportive of all the efforts of me and my company, from intellectual capital development to market insights and on to sales proposals. He was my company's local partner who would be presenting the consulting proposal to PetroFuerte later that day. Juan Alejandro was fifty-five years old, dark-skinned and brown-eyed. He was short and balding in the back, and engaged with you by looking directly in your eyes when he spoke. He took good care of himself physically—he was svelte. He was originally from Venezuela, but lived in Columbia almost all his adult life. He himself was the second of six children, and he and his wife had six children also. Curiously, all six were petroleum engineers, and five of the six work in the consulting business there in Bogotá with him (the oddball brother lives in the United States!). Juan Alejandro got his graduate degree in petroleum engineering at Pennsylvania State University and the University of Missouri, and he spoke English perfectly, but his thick Hispanic accent was overwhelming. Listening to him required extra focus and attention. He was extremely polite, reserved, and demure, perhaps out of the knowledge that his accent made communication a bit challenging. He was very diplomatic, would never offend. He even hesitated before speaking, which came across as a slight stutter.

In addition to representing my company, Juan Alejandro was our technical expert. He and his four brothers in his family consulting and engineering business were all experts in enhanced oil recovery techniques, which makes sense given that all the oil in Venezuela and Colombia, where he lived most of his life, is extremely viscous. He knew every trick about how to make thick oil thin, based on twenty-five years of study and applied science working in Venezuela and Colombia in oil production and oil reservoir engineering, including in global oilfield service companies like Halliburton and the biggest Latin American oil companies in Brazil, Colombia, Venezuela, and Mexico.

Before the meeting with PetroFuerte, Juan Alejandro, Camila Gabriela Moreno Sánchez, a young consultant who worked for me in Bogotá, and I stopped at a café to burn some time. Juan Alejandro parked the car and paid off the car-boy, a lanky thirteen-year-old who was polite and nice, and protected the car while we were in the café. Actually, he would have slashed the tires and scratched the car if we had not slipped him a dollar or two; "protection" is a cover story to dignify the transaction so he does not have to beg. The experience reminded me of the more professional varieties of payoffs like this. While consulting in Brazil, I was driven to work each day in a bullet-proof car because if certain people didn't get their payoffs, heads would roll.

Sipping coffee at the café, appreciating the robust flavor, we got into a long discussion about the renowned coffee industry in Colombia, and how coffee gets to market. Juan Alejandro's cousins were growers, so he was quite knowledgeable, and he offered to introduce me to them. His offer was generous, and I considered it while I excused myself to use the bathroom. The bathroom excursion took longer than I expected because I needed to buy a few sheets of toilet paper from a guy in the bathroom, and I didn't have any coins, so I needed to take an extended tour of the shops in the mall asking people for change for bills, then return to the bathroom again before rejoining my party.

Our buffer time was shrinking and it was time to head to PetroFuerte. Because of dense and unpredictable traffic in Bogotá, we needed to leave ninety minutes before the meeting. Traffic is a problem throughout Latin America. Once it took me over two hours to get from the airport to downtown Mexico City, most of it sitting in stopped traffic. Mexico City was the worst for traffic—you could only schedule one meeting per day because getting around the city was so time-consuming. Much of the traffic problem stems from a lack of public transit systems. In Bogotá there are buses, but no subways or monorails or any other mode of transport that got people off the road, because the city became overbuilt before the idea of public transport came about, and by then the city was so overgrown that it was inconceivable to secure the needed right-of-way from the thousands of property holders in the way.

Juan Alejandro got his "other car" because today was an even day, and due to the pollution, access to the city is restricted to even-numbered license plates on some days and odd-numbered license plates on the other days. To get around the restriction, Juan Alejandro had two cars, one with an odd-numbered license plate and a second with the even-numbered license plate. This defeated the purpose of the pollution controls, but it worked for us.

The route was filled with many small car and tire repair shops and single-floor "road food" type canteens. The streets were a mix of boulevards, access roads, and narrow streets. There were narrow frontages and there wasn't much distinction between commercial and residential buildings.

Our meeting at PetroFuerte went well. Juan Alejandro went back to his office, and he let Camila Gabriela and I take a taxi back to my hotel. We tried to phone ahead for a taxi (this was before ridesharing services like Uber) because that would be more secure—the taxi drivers were a never-ending source of risk. Overcharging and rendering the wrong change was commonplace. Frequently the drivers didn't carry, or pretended not to carry, change because the lack of change inevitably resulted in larger tips for them. One taxi driver faked an elaborate transaction of making change in which he exchanged all of the customer's real money for all of the taxi driver's counterfeit money. Some would drive off the route to rob you. Well, there were no cars available through the taxi service, so we decided to take a taxi, despite the risks.

As we stood in the street to catch a cab, Camila Gabriela coached me. "Keep your phone in your pocket. Don't check messages in the street. Somebody might grab it from your hand." "Keep moving, don't just stand in one place—it makes you a harder target to rob." Indeed, one of the travel advisories I read about said to "walk with purpose." I wore a dark wool suit and a white starched shirt—I think my clothes and movements telegraphed my identity as an American businessman. Other travelers advised to keep things in front of you instead of at your side, and to text the license plate number of any taxi to friends before you get in "just in case." On a travel website, one woman advised females to assign a friend to watch their drinks when they went to the bathroom to avoid being drugged and abused by men.

DISCOVERING COFFEE

Since I had built in some extra time to my trip to Colombia, and there were no pressing meetings to fill it, I took Juan Alejandro up on his offer to introduce me to his coffee grower friend so I could better understand the process.

The next morning Juan Alejandro drove me to the countryside and introduced me to the coffee plantation owner's house. Ana Rosa was twenty-seven years old, short with long dark brown hair and brown eyes. She was wearing jeans and a T-shirt. Her parents were divorced and she had a younger sister who was twenty years younger than she (literally a little sister!), to whom she was more like a mother to her than a sister. A grandmother was holding an infant. I asked about the father, but the answer was not clear to me; it seemed like she said something about her boss.

Ana Rosa's house was set back less than two meters from the highway. Standing in her living room I felt the wind from a passing truck. She walked to work, which normally took forty minutes but today Camila Gabriela could drive us in her car. On the left there was a shantytown. I smelled burning trash—there was no garbage collection or municipal incineration—and spotted the location of the fire by the plume of smoke.

She explained the coffee growing process to us carefully as we toured the farm. I had some trouble understanding her English, but was able to look around and get visual cues about what she was saying, which made communication easier.

Itinerant farmers picked coffee beans in Colombia, working fifteen-hour days during harvest season. They earned the lowest wages of all the occupations in the country. The law provided them little protection, and labor policies were neither posted nor observed. Latin America was less egalitarian and progressive than North America, including Mexico. Females earned 20 percent less than males. The business was run by the men of the family; none of the women held any ownership of the company.

Trucks hauled the beans from the plantation to the roaster in near Monterrey, Mexico. I agreed to take a ride with one of the drivers who was collecting beans from various farms around the region. I could tell from the sound of the engine and the trail of exhaust that the truck badly needed a tune-up. The windshield was cracked on a diagonal across the entire glass. Even if the glass itself was safety glass, that crack would have been very dangerous in a high-speed collision. The roads were rutty and poorly signed and signaled, but the poor signage didn't affect this driver, who obviously knew his route, and luckily the police never saw or enforced the emissions violations. There weren't enough police to notice it anyway, and even if there were, they would not penalize the driver or the owner for motor vehicle violations. If necessary, a small bribe would take care of that. However, the bumps and ditches were risky since the cargo was overloaded and bags of coffee were dangling precariously off the top of the pile. There were no straps or bars to contain the cargo, and if one or two bags fell out from the wrong places, most of the load could fall off the truck in a giant "coffee-slide."

The driver carried a gun in case of an attempted robbery. Crime and corruption were pervasive in Latin America; the region ranked 83 of 168 in Transparency International's Corruption Perceptions Index, next to Liberia.[1] While I was attending a conference of about 120 people in Mexico City, twelve of the attendees' laptops disappeared during a fifteen-minute coffee break. Nobody had even left the room. The perpetrators carefully stole away with each laptop while the owners were looking the other way, including removing a number of them from briefcases and backpacks. Not all the crime was petty: the FARC (Colombian Revolutionary Armed Forces) and the ELN (the National Liberation Army) attacked the oil and gas pipelines 259 times that year. Conventional physical barricades and patrolling were insufficient. The pipelines were secured by many layers of physical and electronic security.[2]

Ana Rosa asked if I wanted to see the roaster operation in Monterrey. At first I thought this would be unnecessary, since this study was just a curiosity, after all. But since my flight to Detroit the next day was connecting in Monterrey anyway, I agreed to take a look. Ana Rosa had been a great hostess and tour guide, and I really appreciated her time and hospitality. I decided to "re-gift" the nice desk clock I had received for speaking at the conference, which was already nicely gift-wrapped. It was the least I could do, and it lightened my load.

In Monterrey I exited the airport instead of following the signs to "Connecting Flights." The roaster was not far from the airport, which allowed me to do this on a layover. Often in my career I have toured factories of various kinds. I always wished that I could bring my wife and kids along on these tours because they are such fascinating learning experiences. As part of my work I have toured an orange processing plant in Tunisia, a caramel candy manufacturing plant in the United States, a windshield manufacturing plant in France, and a container shipping port in Egypt. I have descended a mile underground in a copper mine in Poland, walked the floor of automotive assembly plants in Brazil and Germany, and ridden in the locomotive of a pig-hauling train in China. That day I was going to learn about coffee roasting operations.

I went through the habitual, by now instinctual, process of mapping the process flow, in my head, and on my notepad, as I toured the operation. What I noticed more than the process, however, was the conditions. The floors were dirt, and the machines, which were three generations old, were leaking oil directly onto the dirt. The workers were all women. There was no protective or hygienic clothing. To cap it off, the manager told me that the company was an exclusive offshored arrangement with Koffie King, one of the largest coffee companies in the world, but nobody knows that because it operates under a different name. Koffie King was the only customer, and the company directed everything about the plant, but its name did not appear on any public document, including the company's annual report or financial disclosures.

This led me to investigate health, safety, and environmental regulations in Latin America. Water standards existed, but they varied immensely between countries. As a general rule, wastewater discharge standards in Latin America were less stringent than those in the EU. Colombia and Venezuela had the lowest standards for suspended solids Biochemical Oxygen Demand, and Chemical Oxygen Demand, in all of Latin America.

I rode in the cab of the truck that carries sacks of roasted coffee from the mill to the railhead in Laredo, Texas, where it is loaded onto railcars for shipment to US retailers. A heavyset thirty-three-year-old Mexican man named Pedro drove the truck. Pedro was the second of five children, and took care of his younger siblings as a father would. He was friendly but serious. Pedro does not have a legal driver's license. He had a slight stutter out of insecurity, but a heart of gold, always giving whether he gets it back or not. The truck chugged up the mountain hills to the rail/truck meeting point, leaving a trail of swirling carbon in its wake. Bicyclists fell in behind, breathing in clouds of carbon during their morning exercise. We passed a polyethylene plant. I noticed bottles and cans strewn along the side of the road, and even food refuse, especially food and food packaging trash.

When we arrived in Laredo, a large, friendly man greeted me. Rob was about forty-three, with big eyes and round glasses that made them seem even larger. He smiled a lot, and his body language was overwhelmingly positive—he leaned toward you when he talked—almost awkwardly close—and he made direct eye contact. Not only was he friendly and personable, but Rob was truly a problem solver—a solution guy. All of this made him a consummate salesman of transport and logistics solutions for produce within the Americas.

Rob was the distributor of the coffee within the United States. He arranged logistics for the perishable food trade between Latin American origins and US destinations. Bananas, papayas, oranges, lettuce, snap peas—you name it. He was a freight forwarder who arranged for trucking to a port, cargo shipping to another port, rail haulage if it was a long haul, and more trucking to the retail store. He knew every logistics and transport salesman in every mode and traffic lane, so no matter what you wanted to ship where, he could arrange it (as long as it was a food product).

He was a deep thinker rather than a quick wit. Whereas a lot of other distributors, traders, and supply chain solutions salesman talked fast and promised the world, Rob would cogitate over things. He spoke slowly and was definitely the slowest driver on the road. But he had really good ideas that integrated solutions for all facets of the problem. Rob had a wife and three children, spread from nineteen years old to eight years old, which I got to know well over the ensuing years. Over the next ten years, Rob would go on to be my friend, colleague, and employee.

At the railhead, we were joined by Steve. Steve was the intermodal marketing manager for a large railroad in the United States. He was responsible for helping customers move product long distances, which inevitably required shifting the load between modes of transport—ocean container ship, railroad, and truck. Any combination of these came under his responsibility. His job was to quote end-to-end rates and close the deal with the shipper (the originator of the cargo). After the deal was closed, he handed off the execution of the shipment to the operations team.

Steve was a tall, pale-faced young man with short, blonde hair. He wore button-down oxford shirts with a polo insignia on the breast pocket. He played tennis twice a week. Steve came from money, and he flourished where money was flowing. Money was flowing in intermodal rail transportation during the 1990s through 2010 in the United States. The great economic boom of the 1990s resulted in an unprecedented amount of freight traffic, mostly across the country eastward as Chinese goods flowed in to WalMart and hundreds of other retail chains throughout the country. The rails also got a boost from the North American Free Trade Agreement, which came into effect in 1994. Automotive parts flowed northward from Mexico to Detroit, lowering automakers' costs and giving the automotive industry a temporary reprieve from the persistent and long-term economic pressures exerted by Japanese and Korean competition.

Steve usually made sure the shipment got from the railhead in Laredo to the railhead in the suburbs of Detroit. However, this week Steve had some personal distractions—his eight-year-old son was in soccer finals, so he was attending the games instead of going to the office in the afternoons—so the rail shipment got there on Wednesday instead of Tuesday. The communication of the details of each new customer and new shipment was not fully communicated to the operations people, and there were some omissions and inversions of individual shipments, which delayed receipt of the manifest in Detroit.

The tardiness of this particular rail load annoyed Patrick. Patrick was the dispatcher and manager of ramp operations at Union Delivery Service (UDS). The ramp was the local distribution center for small packages. In this role, he controls the fleet of seventy-two brown delivery trucks that carry the coffee, among many other products, around the region. UDS is expert at picking up and delivering millions of packages every day, and Patrick, a fifteen-year veteran of the company, knew more than anybody about getting it done. To the grocery stores that were waiting for the shipment to replenish their empty shelves, Patrick was UDS.

When all went as planned, Patrick was a jolly man—like Santa Claus without the beard and the hat. He was a large man with a big head and smiling face, always wanting to do good and make his customers happy. He had curly brown hair and brown eyes, and wore functional work clothes every day of the week, workdays and weekends. When shipments were late, the stores blamed it on Patrick. They graded Patrick on their satisfaction, so Patrick's scores fell that week, and since his compensation was based on customer satisfaction, his compensation took a hit.

But it wasn't the money that bothered Patrick. He took excellence personally. Patrick was a linear, black and white thinker, which he probably absorbed during his years of living in France as a young man. He reasoned in a bottom-up, granular way. Would it be a good day at work? He couldn't answer that question without thinking through every one of the one hundred thousand packages that were going to be delivered that day, and estimating how many were going to be delivered without error or delay. To figure out something like how much additional time would be incurred by adding a requirement to get his signature at every stop, most people ballpark the answer based on the total time (for example, fifty trucks times ten hours equals five hundred hours), assume a percent increase like 1 percent, and multiply five hundred hours by 1 percent to get five hours of additional time on the road. It was an estimate, and they would have it within a few minutes. Patrick, on the other hand, would reason from the bottom upward. He would pick one sample shipment, measure its actual impact, repeat the exercise for each unique type of stop—maybe fifty different types of stops—and then you would multiply the actual percent impact on each of the fifty different types of stops times two thousand stops for each type of route (50 types of routes times 2,000 deliveries per type of route = 100,000 packages). It would take Patrick a long time to figure this out, but management was able to take his answer to the bank. Patrick was an industrial engineer.

I respected Patrick's way of thinking, even though it was different from my own. Patrick became a business partner and friend over the years. Our debates strayed far from transportation, ranging from economics to politics and literature. To my surprise, Patrick wrote a book on French social history during the time of Robespierre leading up to the French Revolution. Not what I expected from an industrial engineer.

To Patrick, information technology was central to reliability and speed. He counted on UDS's network of handheld devices to feed real-time information on deliveries to him so he could dispatch drivers dynamically and keep his fleet busy but not overloaded. He participated in pilot programs that integrated that real-time information into real-time traffic information, and had volunteered to be the first ramp to operate all self-driving vehicles. He frequently talked about the infrastructure gap between the United States and other countries. For as large as it is, the United States invested relatively little in smart transport. The federal government had very tight purse strings, and left states and research institutions to pick up most of the cost of research, development, and even early stage implementation. The Transportation Research Board, part of the National Academies of Sciences, Engineering, and Medicine, lived off meager grants from the government, which each required a thorough and bureaucratic application and review process. The United States lagged in real-time traffic information, vehicle-to-infrastructure and vehicle-to-vehicle integration, adoption of computerized traffic signals, and maximizing the effectiveness of its already fielded intelligent transportation systems.

I hopped a taxi back to Monterrey to fly to Detroit. In the plane, I was seated in the aisle seat next to an Asian man who looked out the window and kept to himself, so much that he even had his head in his hand and I could not see his eyes. He smelled, though, of body odor and alcohol, which was unusual, especially in business class. Figuring that we had a nine-hour flight together, I tried to "break the ice" shortly after takeoff with some small talk. He only managed a few words before he broke into tears. Although his English was limited, I was able to understand that he was a Korean ship captain whose vessel had been confiscated while sailing without proper documentation in Brazilian waters. He was being sent home on the first available seat, which happened to be in business class, via Bogotá, and right next to me! He sobbed for most of the rest of the trip, and we didn't talk much more. So, although crime and corruption are rampant, some people get caught, seemingly randomly, and the consequences can be painful.

In Pontiac, Michigan, outside of Detroit, I paid a quick visit to my old friend Luuk Van den Berg, who was the perfect person to fill in the remaining piece of the "coffee supply chain" puzzle. Based on a brief phone discussion, we decided to meet at a Dunkin' Donuts in a strip mall in Auburn Hills. I arrived early and looked around. Most people carried an extra thirty pounds or more, and wore stress on their faces.

Luuk ran the American coffee and tea subsidiary of Koffie King, one of the largest consumer food products companies in the world. Not many people had heard of this company, especially outside of Europe, mostly because it sold coffee under a large number of different brands. Koffie King hired Luuk to increase coffee sales in the United States. Luuk was healthy, prosperous, and smiling. He was a towering 6′8″ Dutch white man who looked big and thought bigger. His imposing presence was appropriate and useful for a CEO. Luuk had a wife, four young children, and a blind old dog at his temporary home in the United States during a five-year expatriate assignment from Holland. The crew was happy. They were flexible and adaptive, spoke English well, and it was their dream to live in the United States like this. Like many Dutch people, he had a quick wit and was a fun-loving, clever, and mischievous oddball who got a thrill out of waking up every day.

We bought two cups of ice coffee and savored the flavor that had made its long journey up from Colombia. I prefer my coffee black, but Luuk ordered the mocha macchiato twist, which had whipped cream and a shot of flavored liqueur in it. We walked away from the cash register with a pile of stuff: two oversized plastic cups, six plastic creamer packets, two each plastic stirrers and spoons, sugar packets, and a fixture that kept four cups standing up straight, for riding in the car (which we weren't doing). We enjoyed the coffee and then threw all the stuff into the trash can. There was no recycling option.

I asked Luuk whether the coffee shop customers are aware of the extensive journey that the coffee makes to get in those cups, about the working conditions in Colombia, the low wages and inequality in Colombia, and about the trucks and the pollution that it creates. After all, American consumers could buy Lebanese coffee that is milled in Spain instead of Colombian coffee that is milled in Mexico, and delivered by US truck drivers instead of Mexican ones. This would alleviate some of the social inequities of farming in Colombia, increase worker safety in the mills, and use more fuel-efficient delivery vehicles. Luuk replied that buying Arabica coffee from Lebanon would increase the price of the cup of coffee by over 20 percent due to the higher price of Lebanese farm workers and the high cost of social protections in Spain, as well as to thousands more miles of ocean and inland transportation, and many transshipments and non-value-added warehousing steps along the way. It would also put farm and factory workers and management professionals out of work in Colombia and Mexico. He explained that the market segment that would prefer to pay extra for environmentally and socially friendly coffee is too small to be attractive to coffee retailers.

I asked Luuk if anyone had considered eliminating the supply chain externalities by reducing consumption, maybe by replacing coffee with tea, which could be grown in the United States. Luuk thought for a second, then replied that tea is a growing market and has great potential. I could see Luuk's mind at work. He had extraordinary powers of conceptualization, and of realization. His mind excelled at business planning, and his charisma and persistence succeeded in getting people to change their deeply embedded habits.

I asked why people consumed so much plastic in the process of drinking coffee. Luuk was friendly and polite as always, but was losing some patience with this inquisition. "Listen," he said. "Everybody wins with coffee. The coffee growers make money. The retailers make money. And Colombian and American economy grows. And as long as Colombian GDP[3] grows, the country gets more loans from the IMF."[4] As long as the money flowed, nobody was concerned with working conditions, quality of life or safety far away, pollution, or trash going into landfills. It was all subordinate to the wheels of commerce. I remembered a water slide I took at an amusement park ride in Costa Rica once. The path was very long and the speed was dangerously fast. There were no safety protections. I burned my butt and could have gotten seriously hurt. However, I didn't get hurt, at least not that time. The resort made money and I had fun. Why change?

I sent my trip report to Alex, my analyst. Alex was a young man from Montana with wavy light brown hair, blue eyes, and a low baritone voice. There was always a glint of mischief in his eye, as though he was up to something. He wore jeans and always looked slightly unkempt: the belt was a little too loose, the pants were a little too long, and the hair needed a little combing. He was rough and rugged. If he were born two hundred years earlier, he would have toted a rifle and a coonskin hat, and his wife and family could have counted on him to hunt, skin the animal, and put dinner on the table in the log cabin. He was a hard worker, and he compensated for his colleagues' weaknesses, which was great because the work always got done no matter what!

I asked Alex to quantify the negative externalities of the coffee supply chain. He came up with a method where he decomposed the "value added" at each supply chain stage and multiplied that value by relative coefficients representing the Depletion, Devastation, and Dehumanization of that industry and that country. He used the inverses of the Global Green Economy Index, the Environmental Performance Index, and the Social Progress Indexes as proxies for Depletion, Devastation, and Dehumanization. The absolute value of any individual resulting number (he called them "D-Factors") was not bounded; there were no maximum or minimum values. Theoretically D-Factors could even be negative if the relevant countries and industries were "energy net positive," he explained to me. However, D-Factors were meaningful when compared to each other: a lower D-Factor was always better than a higher D-Factor.

Later that day, as I parked the rental car and started walking toward my client meeting, an email came across on my phone with Alex's report on the D-Factor. He said that the total D-Factor for the coffee supply chain was 1.8, and spat out some more numbers: 0.4 for Depletion, 0.7 for Devastation, and 0.7 for Dehumanization. I asked him what that meant. He said that the Devastation number represented the loss of natural green landscape from the port cities and the railroad lines, the pavement and loading docks for trucks, and the commercial developments that support this international supply chain. Colombia's rich biodiversity and ecosystems were coming under pressure from extractive industries, livestock grazing, road traffic, and urbanization.[5] Excessive cattle farming was prematurely wearing down farm land. His Dehumanization figure stemmed from the social and financial mistreatment of the farmers in Colombia (wide disparities in income and landholdings are deeply embedded), and poor working conditions in the roaster in Mexico, where the machines posed safety hazards and leached chemicals into the ground below them. The extensive supply chain depleted natural resources through multimodal (truck and ship) transport across 3,000 miles from the farm to Dunkin Donuts in Detroit. His Depletion number was based on the fact that the Colombian land suffered from deforestation and inefficient oil and gas exploration and production.

Back in Detroit, my client meeting was in an old automotive stamping plant. The place was cavernous. It was built during the 1950s, when the manufacturer was producing millions of cars each year. The hallways were made for cars and trucks to drive through. Consulting to this client was tricky because the company was mired in management-labor disputes for decades. The rigidity of the union rules, placed against a dynamic external competitive environment that was crushing them, was pathetic and in the end catastrophic to the company's existence. As consultants, management coached us to avoid using the words "change" and "improvement." The best way to improve efficiency was actually to get rid of excess people, because running the operations with far too many people made it hard to engage in process optimization and to even understand how to correctly run such an operation. The laid-off employees received full salary for life, so downsizing didn't save money, but it allowed a path toward improvement. As the laid-off employees aged, some died and their relatives kept collecting and cashing their paychecks rather than tell the company they had died. Although efficiency could be improved instantly by turning the dial on the stamping machines so they ran faster, union rules disallowed any such approaches. In the end, the company only permitted actions that were "too little, too late."

The world had its share of problems in 2010.

NOTES

1. Transparency International (2015), Corruption Perception Index, www.transparency.org/cpi2015/. Last accessed June 14, 2017.
2. David Jacoby, Pipeline and Oilfield Security Technologies: Global Benchmarks. April 28th, 2014. Presentation to the 5th Colombia Oil & Gas Conference & Exhibition.
3. Gross Domestic Product.
4. International Monetary Fund.
5. OECD (214), *Environmental Performance Reviews: Colombia 2014*, OECD Publishing, Paris.

TWO

Cartagena to New York

BOGOTÁ, APRIL 18, 2030

I land in the Bogotá airport. I have another meeting with PetroFuerte, this time about costs at one of their refineries. The meeting is in Cartagena, where PetroFuerte has built a state-of-the-art refinery. My company is helping to resolve some claims that the construction firm had overcharged for some aspects of building the refinery.

After baggage claim I make my way through the various security checkpoints, toting three bags on this trip instead of the usual two. There are several checkpoints. The first is to ensure that you have your immigration papers. The second is to scan your bags and make sure there is no contraband material in them. The third is a very secure and automated door to exit the security area. Finally, I exit another door to get out of the airport. After passing the security scan and exiting the first (secure) door and the second (unsecured) are to exit the airport, I load my bags in the taxi and we start on our way to the hotel.

About ten minutes into the taxi ride, I realize that I only placed two bags in the taxi, not three. What happened to the third? That one contains my laptop and all sorts of credit cards and banking tokens that would be extremely hard and time-consuming to replace. I rack my brain trying to remember. I definitely put it through the scanner. I definitely did not put in the taxi. That means that it was somewhere between the secure door and the taxi line. Certain that this bag was picked up by criminals crawling around the Bogotá airport due to the long-standing drug trafficking problems there, I instruct the driver to turn around and return to the airport so I can try to recover my bag at the airport.

I search the vestibule between the security door and the exit door from the airport. No bag there. I search the taxi line and ask a number of people if they saw anything. Nothing. Surely my bag had been stolen. I track down the airport police and begin to explain what happened. The first security person introduces me to a second security person at the other end of the airport. I explain in English. He half understands and communicates back to me in somewhat fractured English. The distance, the time, and the language barrier further convince me that the bag is gone forever. After about forty-five minutes of explaining my case to different people, I get a call from Camila Gabriela, who has also been tracking the situation by phone with the various airport authorities. The airport security team has picked up the bag and put it in a secure room.

"No passenger at the airport would pick up your bag," Camila Gabriela explains to me. Due to the drug trade, there is intense security coverage at the airport, and anybody caught in even minor incidents is punished to the maximum extent of the law as part of the crackdown on illegal drug traffickers. Consequently, and ironically, Bogotá is the safest place to leave an important bag lying around. It takes me about seven hours to get my bag back, due to the shift changes and opening and closing hours of the various back office rooms, but the experience gives me confidence in a new and better Colombia than I knew in the past.

REDISCOVERING COFFEE

I land in Bogotá and meet with Juan Alejandro again. We have coffee, and during the meeting we reminisce about my coffee supply chain trip from 2010. A wide array of regulatory changes have occurred since the last visit, which affect the coffee supply chain significantly, so I decide to pay a visit to Ana Rosa again to see how her life and work has changed.

On the way to Ana Rosa's coffee plantation, I notice how beautiful and quaint Cartagena is; it is like a small French or Italian town from the Middle Ages with a fortress, like Aix-en-Provence or Carcassonne in France, or Siena in Italy. There is a nice café and biergarten, and a town square with street performers. Music from a jazz bar floats through the street.

There are people jogging in the park, which reminds me of how I used to jog in the Parque Ibirapuera in Sao Paulo when I lived in Brazil. There is what appears to be a parade, but when I look more closely it is a strike against management of a retail chain. Despite the act of defiance, the event is civil and there is no hostility or violence.

Ana Rosa had two more children since our last visit; now she has three instead of one. She sold her house to the state when they widened the road, and she moved to a housing development away from the highway. There is a park in the middle for the kids. The city built a light rail link, which shortened her trip to work to ten minutes. Overall, a big improvement.

The company adopted the UN Global Compact's ten principles for working conditions, which are: (1) Respect for human rights; (2) No forced labor; (3) No child labor; (4) No employment discrimination; (5) No restrictions on association or collective bargaining; (6) Adequate occupational safety (emergency preparedness, fire safety, industrial hygiene); (7) Precautionary approach to environmental impact; (8) Development of environmentally friendly technologies; (9) No corruption, bribery, and kickbacks, including gifts; and (10) Proper accounting and business records.

Because of this, Ana Rosa has a new boss (who does not harass her), and she has structured working hours and days off. In addition, Koffie King discloses its relationship with the Colombian growing company publicly in its annual report, so Ana Rosa's voice can be heard in the event of future unethical behavior.

Colombia began tracking the Genuine Progress Indicator in addition to Gross Domestic Product. This revealed underlying wage inequality and the size of the underground economy. These revelations triggered political action, which resulted in a more progressive taxation regime and the move toward a paperless economy with proper accounting records. Tax revenue increased, which has provided more money for public services and reduced the cost of electricity and water, which reduced irrigation costs.

The plantation now receives its power from a geothermal power plant, and the old coal-fired thermal plant was closed as a result of a carbon tax that was applied five years ago and is progressively increasing over the next ten years.

Tourism has increased, which has been helped by a reduction in crime. Tourists come to raft in Level 4 rapids and to see the magnificent waterfalls. There is tasty and safe-to-eat seafood based on clean estuaries and beaches. The coastline has become a haven for windsurfers.

In the in the intervening years, the local agent's job has been replaced by the Internet. International freight forwarders set up e-commerce websites, and more and more of the growers got Internet access. Now coffee growers sell their product on auction sites instead of to the agent. They were initially very concerned that auctions would drive the price of their product down. However, they quickly found out that Internet auctions allowed them access to far more buyers than they had previously. They were no longer beholden to individual agents. In fact, buyers from Asia came online and bought their product at higher prices than ever before. Also, the growers established direct relationships with freight forwarders, which eliminated the need for local distributors. The freight forwarder aggregates all the shipments together.

When it was clear that there was no more need for local distributors, Rob went to work for the freight forwarder. He works in a nicer office, drives a nicer car, and still has interaction with his old customers frequently. Although the growers lost the sense of personal connection that they had with Rob, their business ran more smoothly because there were no more delays or errors in the transmission of their shipment orders.

Pedro still drives the same route, but his truck is inspected and meets emissions standards because it is owned by an international company with shareholders and a reputation to uphold.

Clean air legislation and clean fuel mandates have produced cleaner air, cleaner water, and a vibrant beach resort nearby. We go to the beach resort and enjoy lunch in a chic bar-restaurant serving upscale fusion cuisine.

Overall, the town's standard of living has increased. There is beautiful architecture, and glass and steel office buildings, and many four to five five-story apartment buildings have replaced the simple single floor tenement houses that existed there before. Consistent with the increase in standard of living and quality of life, the drug trade had been wiped out by police efforts over the last five to ten years.

We drive up the beautiful, fog covered mountains in a clean vehicle which leaves no exhaust trail behind it. The country has adopted the US air quality standard of 15 μg/m3 per year and a twenty-four-hour standard of 35 μg/m3, one of the healthiest levels in the world.

Luuk has moved to New York. Things have improved there as well, he tells me. He boasts of the increase in his staff skills, which have improved the customer experience, which has in turn increased revenue. The United States is the strongest country in the world for skills development, skills versus job matching, and sustainability of higher wages, according to Hays's Global Skills Index. It scores 9, compared to a global mean value of 7, and Colombia and Mexico's scores are also above the mean.[1]

Luuk's staff has become more expensive in New York than it had in Detroit, since New York adopted a $30/hour minimum wage. New York and other states even protect the rights of stay-at-home workers.[2]

At a café on Lexington Avenue we buy coffee. It comes in completely recyclable material. There is no plastic, and the store has a recycling bin. The menu has D-Factor labels with information about the origin of the coffee and which provide an index that has been calculated by an objective non-governmental organization that indicates how ecologically and socially progressive the supply chain for that product is. I choose the coffee from northern Colombia instead of southern Colombia because of its lower D-Factor.

As I sip the coffee, I appreciate the clean air. The US Environmental Protection Agency (EPA) has established goals for each state in terms of a rate-based state goal measured in pounds per megawatt hour (lb/MWh), a mass-based state goal measured in total short tons of CO_2, and a mass-based state goal with a new source complement measured in total short tons of CO_2. States are responsible for implementing the standards and utilities for meeting them. Each state has deployed an assortment of policies to lower its emissions, such as the retirement of aging power plants and fuel switching at existing plants; the optimization of power plant efficiency, heat rates, and dispatching order; and the implementation of Combined Heat and Power technology. Power plants have improved coal quality, increased generation from low-emission resources, deployed carbon capture sequestration, reduced losses in the transmission and distribution system, and increased the accessibility of demand response systems.

The US's Clean Power Plan set a national limit on carbon pollution produced from power plants. The plan is reducing carbon dioxide emissions from electrical power generation by 32 percent within twenty-five years relative to 2005 levels. The plan has reduced the pollutants that contribute to smog and soot by 25 percent, and the reduction led to net climate and health benefits of $45 billion per year.[3]

Additional revenue from a fossil fuel tax has funded refurbishment of all the museums and public parks in the city. Many old public housing developments have been knocked down and rebuilt with new technology, including energy-efficient buildings. In addition, the city now borrows money for its capital projects by issuing green bonds. These are dedicated to projects that are uniquely sustainable in nature, which constitute a large proportion of the ongoing projects.

I call Alex and ask him to run the D-Factor numbers. How much better has the supply chain for coffee gotten since I toured Ana Rosa's coffee plantation for the first time ten years ago?

Alex takes into account the various factors I described above, and comes back with his analysis. The total D-Factor has shrunk by 26 percent compared to the old supply chain, he says. He explains that Mexico has implemented a fossil fuel tax, which has reduced the amount and the emissions from truck and rail transport, albeit at a reduced rate compared to Canada and the United States. The Depletion factor went down by 29 percent.

Mexico and the United States mandated zoning and right of way laws, which separated the rail and truck lines from residential districts and has enhanced livability of towns on the transport path. Mexico also upgraded the rail and road networks, which has made the packaging completely recyclable, and upgraded the standards for the motors in the milling, which reduced oil leakage and seepage. These changes together have reduced the Devastation impact by 25 percent. Plus, Colombia has implemented and enforced tighter labor standards, and the containership that carries some of the coffee beans (those that aren't sent by rail or truck) to the US coast has been reflagged from Liberia to Mexico, and the workers are Mexican instead of Philippine slaves. Together these two factors have driven the Dehumanization factor down by 25 percent.

The combination of improvements to the environmental and social aspects of the distribution of coffee have reduced the overall average D-Factor for Colombian coffee by 26 percent compared to the value chain we looked at in 2010.

In addition to what happened specifically in Colombia and New York, some changes in the broader international regulatory environment have improved sustainability in both places. The World Trade Organization (WTO) has changed its rules to mandate performance of all members to environmental and social standards promulgated by the United Nations Framework Convention on Climate Change (UNFCCC) and various organizations like the UN Global Compact. Previously, WTO rules prohibited the receiving country from dictating the environmental or social standards that were used to produce the product they are importing. The change drove cleaner global supply chains between Colombia and the United States as part of the compliance with the global trade standard rather than any internal regulations or bilateral agreements. In addition, the United States and Mexico have capped aircraft emissions at 2020 levels, which has incentivized more efficient aircraft and operating rules. If the airlines don't comply with the emissions standard, they have to pay penalties equal to the cost of compliance ("play or pay"), so they all comply.

I depart for Boston feeling pretty good about the world.

NOTES

1. "Labour Markets in a World of Continuous Change," The Hays Global Skills Index 2015, p. 7.
2. ABC News, "Illinois Latest State to Expand Domestic Workers Rights," http://abcnews.go.com/US/wireStory/illinois-latest-state-expand-domestic-workers-rights-41553585. Accessed June 14, 2016.
3. "6 Things Every American Should Know About the Clean Power Plan." August 3, 2015. By Gina McCarthy. US Environmental Protection Agency. https://www.epa.gov/newsreleases/6-things-every-american-should-know-about-clean-power-plan

THREE

Detroit to Abuja

LONDON HEATHROW AIRPORT, APRIL 29, 2010

There was an altercation at the gate. One of the passengers got into a loud cursing fight with the boarding attendant. I had started my trip from Detroit and was one of two white people of 524 passengers on the connecting flight from London Heathrow to Lagos, Nigeria, a rare situation for a white American male who is usually in the majority in Western business travel. The flight was fine, but on the other end there was another loud altercation, between the same people!

Walking through the airport in Lagos, agents stopped me for a medical screening to make sure all entering passengers had certain immunizations. I had my immunizations and passed through, but immigration officials were in the process of sending an entire aircraft back to Senegal in retribution for Senegal's rejection of an entire plane load of people the previous day from Nigeria for lack of immunization documentation. It was a political spat between the two countries. The whole thing was bizarre—the equivalent of another altercation, except between nations.

Olufunmilola Babatunde, my company's Nigerian-based analyst and in-country representative, arranged for a driver to pick me up at the airport. "Lola," as I called her, is a "do-gooder." Her career goal is to become the minister of education in Nigeria, and she laid the groundwork well by volunteering in leadership positions in several not-for-profit associations and organizations that help bring people out of poverty. She cofounded and codirected Against Illiteracy Poverty and AIDS (AIPA) in Lagos, where she organized, handled logistics, and managed over one hundred personnel involved in various fundraising events, including a fundraising effort that generated $20,000 and raised AIDS awareness. She was president of the Nigerian chapter of the Society of

Women Engineers, and a Big Sister. Her parents raised Lola right: she was always bringing gifts, small or large. She always had something in hand.

Despite Lola's good intentions, the driver who was scheduled to pick me up never showed up. Maybe the flight was too early for his taste (5:30 AM arrival). I sank into a metal bench, as much as that was possible, at the airport. Lola ended up picking me up with her driver three hours later, at 8:30 AM.

On the way to the hotel, some fake policemen stopped our car, got into it with us, and demanded that we go to the police station for a fabricated traffic violation. Lola had just picked me up at the Lagos airport and we were on the way to my hotel. She instructed the driver to take a U-turn and he tried, but men in uniforms standing there didn't like that. There were lots of police everywhere, and there were more police standing around than there were police cars, and they were dressed like military police. They all look unhappy, and were probably underpaid, underqualified, and underfed. If they were legitimate at all, they seemed to be there for crowd control rather than enforcing rules. Anyway, two of the "policemen" pulled our car over, told our driver to roll down the window, and started yelling at him and leaning into the open windows. Then they told the driver to pull over and said they were going to take him (and me and Lola, apparently) to their office. It was unclear where their office was.

Lola told us to lock all the doors, and a minute later the "policeman" was trying to get into the car. Lola quietly and calmly dialed her dad on her cellphone as the conflict escalated. Then one of the policemen told the driver to unlock the doors and he got into the car and sat in the front seat while the other one continued harassing us from outside the car. Lola passed her cell phone to the guy sitting in the car, telling him that "a man wanted to talk to him." For ten minutes the two policemen passed Lola's cell phone and their own cell phones back and forth, talking to the man on the other end of the line and then calling him back, asking and answering basic questions like their name, where they were based, their badge numbers, etc. Then they released us on the premise that Lola's father was to call them back later in the morning to "finalize things." I wasn't sure what exactly that was to entail—maybe a bribe, or filling out some papers about the incident, or clearing the records of the guys who otherwise might have had black marks in some system, or just a stern talk?

When we were back en route to the hotel, Lola told me that they just wanted to be bribed but she didn't want it to end with a bribe. Her father used a bogus identity and made them feel like they were talking to somebody very important. He never intended to call them back the next morning. The whole event crossed a line—they were acting as police but they weren't protecting us or anybody else. They were harassing us, and it seemed kind of invasive and dangerous. Obviously, native Nigerians have developed defense mechanisms for this.

I was in Nigeria to establish a partnership with an engineering firm, to take advantage of the booming oil and gas industry there. Oil and gas was becoming a huge business during the oil boom of 2007 and 2008, and international oil companies like Chevron and Shell, as well as national oil companies like the one in Nigeria, were investing heavily in offshore production. My company provided a range of studies to determine which projects to pursue and execute them cost-efficiently.

One of my analysts had gone through a thorough vetting process to select a business partner in Nigeria, and he had scheduled a meeting between me and the company's president, Dr. Fasasi Abdoullahi Agboola. Dr. Agboola was a short Nigerian man. He was bald and wore intellectual glasses. The frame was old-fashioned and the lenses were round. Agboola was always well-dressed, whether at the office or on the golf course. He seemed to be a true gentleman. He climbed his way to the upper crust of society, and did not rest on his laurels. He was a bigwig in Nigerian business. He represented Nigeria at OPEC,[1] and had been an executive at the Nigerian National Oil Corporation and in the planning department at the Nigerian Petroleum Inspectorate. He had studied at Oxford University, which was extraordinary for Nigerians.

Agboola always seemed to have everyone do things for him—place phone calls, communicate things to other people, fetch the car, tell jokes, everything. The man just talked, and twice a week swung a golf club, but he seemed to control half the country!

Dr. Agboola and I were helping the Nigerian government develop a strategy to increase "local content" in the oil and gas industry, helping the national oil company establish green carbon credits, and helping one of the largest European multinationals to establish manufacturing, assembly, and repair operations in Nigeria.

Agboola and I thought alike in many ways. He was a straight shooter, said what he thought, and did not bend to others' political wishes just to win projects or enhance his social status. He steered away from corruption, despite the fact that it was all around him and abstaining from it clearly put him in a losing position in many business situations. He thought long-term and kept his integrity and his conscience intact. His soul would outlast the noise and selfishness of the nasty business going on around him, I thought.

It was hard to be so fastidious, and conversely it was tempting to be corrupt, to lie, to bribe, to cheat, to go around people and systems because nothing really worked like it was supposed to in Nigeria. The hotel had an exercise room, but none of the equipment (treadmill, cycle, or weights) worked. There was air conditioning, but it was precarious and very weak. There was internet, but the connection didn't work. There were ATMs, but none of them had money in them (I tried them all!). You had to go into a branch and see a teller to get money.

The nonfunctionality was not limited to things. People said they would do things, and then . . . they seemed to completely forget ever saying they would do them. After one visit to the Nigerian National Petroleum Company, involving a prescheduled appointment and in-person visits, one executive didn't remember me or my visit, at all, when I called back to follow up on something we had discussed. Didn't ring a bell. After scheduling a trip to Ghana to meet a project manager about a refinery project there, the day of our meeting came and went, and despite numerous calls and emails from me, the person never replied to me with a time or place to go. He just never responded, at all, ever again.

Everything was a bit off, in a surreal way. On flights of stairs, none of the steps had the same rise. The joints between walls were not straight (not built with tools?). Building construction was ubiquitous, but most of the buildings were left unfinished, with no tools, equipment, or supplies in sight, just abandoned. Construction supplies like doors were made of old, used, bent metal. Payment for a million-dollar construction crane was made in cash and filtered through a variety of blue-collar middlemen who were involved in every phase of transportation and receipt, each taking their share. I felt like I was living inside a Picasso painting where everything was bent or distorted.

Tiny, old, boxy econo-cars and minibuses were riddled with dings and dents, as if they were made of tinfoil and somebody crumpled the tinfoil as hard as they could and then flattened them out again. I couldn't figure out how they got that way. There were bugs in the hotel room, the toilet seat was too small, and the plastic sink was tiny. The beef was stringy and the bread was inedible. The dining area smelled of burnt kerosene from the stoves, so I sat in the lobby to eat. The stagnant water in gutters and ditches was visibly contaminated. Exhaust belched from trucks and cars. A brushfire next door to the hotel created a forty-foot fire with smoke blowing zero visibility into the left side of the hotel. Nobody said anything about the fire, or about anything else. This was normal.

The average lifespan was 47.5 years. Statistically, I would have already been dead if I lived there. Going to the hospital for any care entailed a high risk of dying from complications because the hospitals had inadequate equipment and supplies. Lola rattled off the names of two young people she knew who died recently from a mundane trip to the hospital. They seemed relatively healthy, but then something happened (asthma, in one case), and they went to the hospital and one thing led to another.

Despite the surreal craziness, the culture was vibrant and the metropolis was thriving. People were carrying anything and everything on their heads. A woman ran across the street with a big vase on her head. A man carried a bundle of twenty-foot-long steel pipes on his head. *Nigerian Idol* was madly popular, and the TV pushed CNN and other cable TV channels to each room. News streamed in from CNN, TF1, and the Middle East. Lola and I had dinner at a chic restaurant and bar, and ended up in a "tequila night" sponsored by Olmeca, a Mexican tequila. We met some friends of Lola, including one who she did her national service with, and there was a radio personality there named Suzo, who had tattoos across his face.

DISCOVERING AIR TRANSPORT

Lola and I had to go to Abuja the next day to meet with a Nigerian petroleum company. As we prepared to get into another airplane, I wondered if there would be more altercations. And, as I still felt weary from the long trip there, my mind wandered to how much impact I was having on the ozone by flying around so much. After all, if I was concerned about the ecological impact of trucks chugging up the hills in Colombia, weren't all these airplane rides doing more damage than that? I decided to find out by learning, while I was moving about to client meetings, more about air transport.

Lola and I talked about my curiosity, and she knew several people who worked for the airlines and in the airport. She made a point of helping me with my special project. On the way through the airport, she introduced me to Victoria Ademola, the ground operations supervisor in the Lagos airport. Victoria had jet black skin, black hair in cornrows, and a round face. She was a very good communicator, engaging directly with us, and smiling at the various ground personnel as we walked by. She knew everybody.

Victoria was involved in a big upgrade project at the airport, so she knew about engineering contracting. "There is a three-tier economy in Nigeria," she explained. "The daily rate for engineering labor is three-tiered, not by skill set but by nationality: $70–$120/hour for Western engineers, $30–70/hour for Asian engineers, and $10–40/hour for Nigerian engineers." The disparity in pay didn't surprise Lola, whose sister was a fashion model. A fashion show was scheduled for the next weekend. Her sister and the other Nigerian models were being paid $200/day, while some of the Western models were taking in about $10,000/day. There was an uprising over the pay differential, which almost resulted in the show being called off.

Victoria accompanied me while I bought my ticket. At first I tried to do it in the main ticket hall, which was swarming with people. The crowd was so thick that I couldn't really see where I was going, and had to continually push on the people in front of me to keep moving forward.

Once I reached the ticket counter, I got good news and bad news. The good news was that plane tickets from Detroit to Abuja were cheap.

There were no income, use, excise, or airport taxes, which kept the ticket prices low. There were no environmental protections, and that meant there was unlimited noise and CO_2 emissions, and limited safety. The week after I returned to the United States, a flight on the same airline and the same route crashed, killing all onboard and hundreds of residents in a suburb of Lagos. The bad news was that I couldn't buy the ticket with a credit card, so I needed to go upstairs to change dollars for Nigerian naira. This involved a free form oral negotiation about the exchange rate. There was no official exchange rate posted. There was no verifiable figure. The rate was not documented on a receipt. Actually, there was no receipt at all. In the end, the money changer just handed me a ticket voucher. I returned downstairs with the voucher and pushed my way through the crowd, which had actually gotten worse than before. I got the ticket, then had to wait in a different line to check my bag, and a third line to get to the gate. The experience took nearly three hours, but Lola and I made it to the boarding gate on time.

Seventeen minutes before the scheduled departure time the airline canceled the flight. Lola had her family's travel agent send him to meet us at the airport and deal with this kind of situation. For an hour, the guy ran around talking to four other airlines to get our tickets exchanged.

When we eventually got to Abuja and stepped outside, the air smelled burnt and looked cloudy. Lola said it was desert air since we were getting closer to the Sahara Desert. She had arranged for the hotel taxi service to pick us up. Upon our arrival she was phoning the driver asking him what he was wearing—not so we could find each other, but to make sure we didn't get abducted by an opportunist picking people up from the airport and kidnapping them for ransom. As we drove to the Ministry of Transport, traffic bottlenecked as cars slowed down and turned to avoid hitting a man with no arms and no legs who was begging in the middle of the highway.

I wanted to know the government's plans for the future, so we paid a visit to the Ministry of Transport. In the lobby, a sign warned us: "Hold Body to Yourself: Wait till Marry." We walked down a long narrow hall past dark wooden offices. It felt like an old schoolhouse, with scratched old wooden desks in tiny rooms, each room marked with a sign on the door. We met in the office labeled Bureau of Rail and Transit, and waited for an hour with many other people in the waiting room. The receptionist watched cartoons on her phone.

When we went into the office, a Chinese man who had been waiting in the office with us also walked in. A big guy in a robe told us to "start." He did not introduce himself or anybody else in the room. I asked who the Chinese guy was. He said he was a friend of the guy we were meeting. This was strange, because after our meeting, the Chinese guy left the whole building, as we did, and grabbed a car. He had been invited to the Ministry just for that meeting. The transport maps and plans were all in oversized hardcopy books. Nothing was available digitally, nor was it posted on any website. There were stacks of reports and files and papers in the office. These were the only copies.

During our greetings, Lola reached out her hand to the men for a handshake, but they declined. This Muslim area in the north of Nigeria was Shariah-compliant. Lola mentioned that a relative of hers high up in the health ministry sent her regards to the potential client. That got a lot of respect and immediately put us in the category of "family" with the guy. Later, Lola told me that her connection at the health ministry never even knew we were meeting with these people; she just said it to establish a personal connection, and it worked.

Although the restaurant for lunch was only five doors away, my hosts would not allow me to walk; I took the car and driver. If walking on any street became necessary, the rule was to keep moving. At lunch, the electricity went off while I waited for the food, but the generators took over and the lights came back on in a few minutes later.

Lola and I drove to the hotel. As we drove by, mechanical cameras snapped pictures of the car from the roadside for security. We spent the night in Abuja. This was Boko Haram country, so a man with a metal detector carefully inspected our whole car before we entered the driveway. The hotel was nice inside, but there were thick iron grates on the windows, and no line of sight to the outside. Good, I thought—a sniper would have a tough time picking me off from the outside. The toilet seat was sitting on the floor and there were buckets and cleaning utensils in the tub. There was no bottled water and the tap water was not potable. The internet signal was too weak to connect and the TV signal was grainy. But the bed was comfortable and I slept like a baby.

The next morning, breakfast at the hotel was good: I decided I could eat fried plantains and coffee for breakfast every day as a habit. Since the hotel would not take my credit card because it was a foreign credit card, I walked to the bank around the corner to get cash. Three of the four ATMs outside were out of order, and the fourth wouldn't accept my card. The bank was well-guarded: there were "police" or military (I couldn't tell) all over. Three lines of people waited to exchange chits for wads of cash, while the other employees were sitting at their desks doing nothing.

We enjoyed driving on a brand new five-lane highway on the ride back to the Abuja airport. Strangely, however, there were no lane markings, and only about one out of every ten massive streetlamps was working. Major luminaires were slanted at twenty-five degrees. Were they put in crooked and just left that way, or did they fall that way over time? Our driver ripped along at 120 kph in the pitch black of night, dodging cars that were parked in the middle of the road and honking at throngs of people running across the highway. Other cars were coming at us, driving the wrong direction. Gangs of cars and motorcycles stopped in clusters in the right lanes and shoulder, randomly. Every 10 km we stopped at "diversions," which slowed traffic down for security checks. As we got off the highway, a much more diverse range of obstacles clogged the road, including tripods (like "tut-tuts" in Thailand), motorcycles, cars, buses, vans, and people hanging out. As we passed the headquarters of one of the international oil companies, I spotted Mercedes, Bentleys, and Range Rovers. Throngs of people hanging out in the middle of the street. Selling things (snacks, chairs, phone cards, anything) to people in cars.

In the airport we met with Omoteleola Ilesanmi, the cargo operations manager, who worked for a company called ShipEX. She was a tall, dark Nigerian woman who carried herself straight and engaged you when she talked. She looked you in the eye and waited for a visual confirmation before moving to the next point. Omoteleola was worldly and well-educated. She was trained in law at the University of Lagos. She had been with ShipEX for many years. One of her daughters lives in the United States, and she visits her frequently. Omoteleola and I subsequently developed a business relationship. My consulting firm helped her company to find an international freight forwarding partner, which is very challenging because when foreign companies hear the word Nigeria, they usually run away. She was persistent and methodical in the negotiations; she kept going for three rounds until she felt that the agreement was ready for a signature. Over the next two years, I introduced Omoteleola to over a dozen potential partners, including two European and Middle Eastern industry leaders.

The flight back to Lagos was delayed, which did not surprise me after the morning's experience. Once we proceeded to the waiting rooms for the gates (which is a big part of the airport), there was nothing to eat or drink. Nothing at all, in fact, except chairs. It was a long wait.

Back in Lagos we needed to drive again in the car from the airport to dinner, which was quite a long drive. On the bridge, beggars knocked on the windows. They came from the nearby marsh-slums, which stretched across the entire horizon. Seventy to ninety percent of youths were unemployed, versus the 15 to 20 percent announced by the government. Some people think that a "cabala" controls the country: rich businessmen who own the big banks and construction companies don't want the country to progress because it would diminish the wealth of their own empires.

The next evening Lola and I had a very nice dinner with Dr. Agboola at an exclusive upscale restaurant. I had pounded yam with ochre soup, which was mushy and slimy and tasted good. The pounded yam looked like a baseball of paste. Lola ate a cow's head (her favorite). We talked about all sorts of business, but I especially remember the discussion about how to bribe officials. Dr. Agboola made it clear at many points during our relationship that he doesn't do business that way, but he is an "old salt," and as any businessman his age would, he clearly knows how it's done. After the dinner, Dr. Agboola paid the bill with a large wad of cash. Lola couldn't finish her cow's head, so she took it home.

Once I retired to the hotel, I decided to focus in on the impact of our flights. I wrote a recap of the trips, focusing on the flights and the airport operations, and emailed it to Alex. I asked Alex what the D-Factor was for the flights we took.

Since Alex was located on the West Coast of the United States, and I was in Nigeria, he worked on this while I slept. The next morning when I opened my email, I read his reply.

Alex said that the total D-Factor was 2,023—the equivalent of 1,123 cups of coffee.

Fuel was the single largest cost component, and flight operations, ground operations, and aircraft and power plant maintenance, were the other large costs.

Using a transatlantic flight as the basis, he calculated a Depletion factor of 1,385. Most of this was from consuming fuel inflight, while a much smaller amount came from consuming oil and gas in ground support operations. Aircraft and airline industry emissions were not regulated at all in Nigeria.[2] The Devastation factor was 494. If takeoff occurred after midnight, this created noise pollution in the communities around the airport. The development of infrastructure in and around the airports paved over wildlife and forced them to move their habitats, many of whom never found suitable alternative homes. The jets created toxic exhaust fumes in the airport area during taxiing, takeoffs, and landings. The Dehumanization factor was 145. Hiring of the flight crew was based on a formalized bias for young, pretty, female stewardesses, who were held to strict weight limits and decorum standards in order to keep their jobs. If they were ever to consider lodging a complaint, they would probably be discouraged by the fact that legal cases were nearly always resolved in favor of the employer.

Passengers had little control over the environmental or social footprint of their flight. Fuel comprised 28 percent of the cost, and the airline's purchasing team strove hard to locate the lowest-cost fuel. A passenger couldn't choose a different fuel for the aircraft, but if he or she could have done so, then switching from Middle Eastern fuel to American fuel could potentially have reduced the D-Factor by 20 percent based on US businesses' more stringent ecological and social institutions, norms, and laws. The passengers also had no control over the aircraft maintenance, but if the airline had chosen to send its maintenance to German vendors instead of Brazilian vendors, that would have shaved another 1 percent off the D-Factor, based on the policy environment of Germany versus that of Brazil. Passengers did have a lot of control over the decision whether or not to fly. "Staying local" could have mitigated the environmental, ecological, and social costs. However, this would have constrained lifestyles, which would have been unlikely to happen in the absence of an economic crisis.

In this case, a more sustainable flight would have cost 24 percent more, so it is doubtful that airlines or passengers would have chosen this path unless they were forced to do so by a fiscal or regulatory mandate.

For my last night in Lagos, Lola's parents invited me to a very special dinner at a restaurant called Red Pepper. The atmosphere and the food were great! We discussed politics and religion, the two things my parents always taught me to avoid in polite social situations. Was there a conspiracy between Great Britain and France to keep the African countries poor and fragmented by preventing the West African Currency Union and the West African Central Bank from becoming truly successful because they were routing banking flows through London? Were the Chinese trying to dominate African economies because the Nigerians were too polite to push back? Would Boko Haram succeed in proselytizing and spreading its influence at any cost? Was Nigeria a pawn in a global economic power shift, reopened anew by America's long decline and the relative reascendance of Asian and European countries? These concerns about Africa being a perpetual colony of various economic imperialists reminded me of my old history lessons.

Back at the hotel, I prepare myself for a very important sales pitch I need to make to a large audience in Saudi Arabia two days from now. Tomorrow I will travel from Lagos to Dammam, Saudi Arabia. I charge my computer and phone batteries and pack for a long flight with a connection in Muscat, Oman.

NOTES

1. Organization of the Petroleum Exporting Countries.
2. T. Streissguth (2011), "Transportation," in *The Role of Industry (Confronting Global Warming)*, Greenhaven Press.

FOUR

New York to Cairo

CAIRO, APRIL 25, 2030

I decide to take a tour of Cairo. Since I haven't educated myself on the city in advance, and since I am alone, I decide to hire a tour guide for the day—somebody who could show me the important touristic highlights and explain it all within the eleven hours I had left before I need to leave for my scheduled outbound flight. The hotel sets me up with Mariam, an outgoing, thirty-five-year-old Egyptian lady. Mariam seems to be acquiring the knowledge of the outside world by meeting international tourists in this way. Her dream is to go to California. She gives me a great tour of all the major sites, with just enough historical detail to be interesting but not so much as to be boring or dull. We see Old Cairo, the Citadel, the Mohammed Ali Mosque, the Pyramids, and the Sphinx. Most interestingly, I get a view of Egyptian history and politics straight from an Egyptian.

Old Cairo is amazing. After going through a security check (where they ask what nationality I am), I visit the place where they took Moses's basket out of the Nile River, and the oldest synagogue in the world, which was converted from a Coptic (Christian) church. Although it is a Jewish treasure, the Egyptians refused to give the synagogue back to Israel, and Israel didn't want the security burden of escorting people to and from the site, so it remains under Egyptian control. There are about one hundred Jews in Egypt, according to Mariam, all very successful and well-connected.

At the Citadel, Mariam gives me a brief Egyptian history lesson as throngs of schoolchildren from the countryside roam around. King Farouq was the last King. The French overthrew Egypt in 1911 and made it a republic. Mohammed Ali was a tyrant, set up the Citadel so he could repress everybody, killed ten thousand people. President Anwar Sadat was assassinated in 1981. President Mubarak succeeded Sadat and survived four elections in what increasingly became a life reign. Under his rule, everybody was too repressed to speak out.

The ceiling of the Mohammed Ali Mosque is an impressive structure, even today, at eighty-two meters high in the center. I saw where the imam leads the prayer. Egypt is Sunni Moslem. The Sunnis don't believe that anyone can take the place of Mohammed's direct lineage. They also, relatedly, don't believe in "alternation," the belief that someone else can do good deeds on your behalf; "you have to be good yourself," explains Mariam. They pray to the East five times each day "to unify the Moslems socially and reinforce the brotherhood of Islam."

The Pyramids were built in 2300 B.C. There are three big ones for the three kings (pyramids of Khufu, Khafre, and Menkaure), and three little ones for Pharaoh Khufu's wives. We only see the major pyramids today; many pyramids were built over a long time. The Sphinx consists of half-man's head and half lion. It represents wisdom and strength, and is credited for guarding the pyramids against invaders (the Crusaders).

Mariam asks me how it was as a Jew visiting an Arab country. I explain that only one of my parents has Jewish roots, and that I am a practicing Christian. Throughout the day, she continues to make references to me being a Jew over and over again. Over lunch, she explains that she is in an arranged marriage with one of her cousins, and that although she respects her husband, she does not love him. She considers herself stupid for still agreeing to do things for him. I suppose she is considering some action like rebelling against his authority or seeking a divorce.

There seem to be no clocks where I would expect them. On the second day I figure out that in my hotel room the TV has an LCD clock, but there is no alarm clock. The Arabic culture doesn't value time as Western cultures do. "Time is money" simply isn't true here. In the Middle East, contract negotiations can take years. The cycle time to get things done is often so long that circumstances change before formalities are concluded, resulting in a big waste of effort and time. For example, political regimes and economic cycles reverse two thirds of the way through contract negotiations, unsettling economic conditions or changing out the sponsors of the program for new people, resulting in the contracting efforts being abandoned. It's part of business and life, and I have gotten better at making judgment calls about when to invest in business deals and when to back away, given this reality.

I am in Cairo to give a presentation to a conference on Cairo's ports. I get an introduction to the Suez Canal Container Authority, and tour the Suez Canal container terminal. At a formal business dinner on the first night polite conversation directly refers to Israel as the enemy. This is not the Gulf Cooperation Council of rich Arabian Gulf nations that is relatively Western in its culture, values, and wealth. It is the Middle East that

Americans learn about in history class. On the second night the conference attendees enjoy a Nile cruise with belly dancing on the stage—a grand extravaganza.

The last time I was here (in 2010), the people I was mingling with had very limited (or no) budgets, and if they had them, they could only spend them in very limited, specific ways. They did not have the liberty—intellectual, political, or economic—to think out-of-the-box or hire consultants for creative ideas. "Ivory tower" consulting was of no help. You had to have very specific information that can be of use to people in their actual situation—like how to optimize specifications for a certain brand of container-moving crane—or you were offering no value. Maybe this was due to the nature of the conference, but I think it was mostly a reflection of the state of the country.

A wide array of changes has occurred since my last visit in 2010. Khalil el-Sisi is no longer in power. The new government has instituted a carbon tax, mandated clean power and ethanol-based petroleum; mandated 100 percent recyclable consumer products; set up a sustainability reporting system that includes GPI instead of GNP, mandated compliance with the UN Global Compact; mandated D-Factor labeling; and participated on a United Nations Industrial Development Organization (UNIDO) working committee for human zoning standards.

I find Mariam on FaceBook and let her know I'm in town. She feels she can talk freely and openly now that the Mubarak and the el-Sisi governments are gone, so we meet at a Starbucks, of which there are many now in Cairo.

There are no more military police on the street corners like there were on my earlier visit. Back then there were police everywhere—military police, tourist police, and traffic police—at least that's what I could identify, and there were probably more branches represented. About a quarter of them were carrying machine guns, and the rest were just standing there. They were attentive and alert, but it was not clear whether they were protecting or ready to attack. Maybe the el-Sisi government had decided that giving tens of thousands of police jobs to people was a good idea for a poverty alleviation program, and also could double as a source of paramilitary power in the event of a coup attempt. Anyway, now the military was separate from the police, and the police just take care of basic street duties like handling mischief and misdemeanors.

The standard of living has increased markedly. In 2010 many cars—Ladas, Skodas, Fiats, tiny Renaults—were still from 1960, and they were all falling apart. Wizened peasants in robes begged for small amounts of money or tried to sell you things in the airport and on the street. Three to five people were hanging around everywhere, seemingly with nothing to do. Since then the middle class has grown, the unemployment rate has fallen, and the currency has become convertible, which has greatly aided exports and tourism. The cars are current models, the shops have been upgraded to boutiques, people are busy, and handheld technology is ubiquitous. Everybody is better dressed and the begging peasants are gone. The people are definitely better off.

REDISCOVERING AIR TRANSPORT

At a lunch sponsored by the conference, I am seated across from Mahmoud, who works for a family owned conglomerate that owns a shipping line, a freight forwarder, a real estate holding company, and a travel agency. This chance lunch with Mahmoud seems like the perfect opportunity for me to refresh the research I did on airport and airline operations in 2010.

His travel agency previously had a network of physical offices throughout Cairo where people used to go to buy paper tickets. The network has been slimmed down to one head office, which is mostly a utilitarian "back office" which customers rarely visit, because everybody buys their tickets online these days. It took more than twenty years, but Egypt has gone the same way as the rest of the world. The travel agents from years ago have found other jobs working in retail stores or service industries like banking.

After giving my presentation to the ports conference, six people from the audience queue up to talk to me about pipeline security technologies such as unmanned aerial surveillance vehicles (drones), radar, Light Detection and Ranging (LIDAR), Infrared Imaging, and Smart Pigging. At the end of the line, waiting patiently, is Raj Balakrishnan, Operations Vice President at Egypt Air. I met Raj years earlier in Monaco while I was giving a speech to a Ground Handling International conference on the use of Radio Frequency Identification technology in baggage handling. The best part of that conference was flying into Monaco on a helicopter at night. Monaco is an enclave of old-fashioned grand hotels and narrow city streets, all on a long strip of densely boutique-filled waterfront. Monaco is tiny, but it is the most densely populated sovereign state in the world. At night, every hotel and boutique is showered in light, which contrasts sharply against the dark backdrop of the Mediterranean Sea in front of the city and the Alps behind it. The view is simply awesome.

I knew when I first met Raj at that conference that we would develop a lasting friendship. From the beginning, he was respectful, humble and courteous, and intellectually curious. Many people of his stature and success would have put the burden on me, as a newcomer, to break my way in, to court him or say something daring or dazzling to impress him, to get his attention. Raj is not like that. Raj is a short, stout Indian man who always wears a suit and tie. He is friendly and effusive, in a businessy way. His "business networking" skill is beyond measure. He is friendly in business relations in the way that you might mingle at a cocktail party with people you don't know very well. He builds trust and relationships in business settings that help to get things done. He believes that a solid human connection is a necessary foundation for any business relationship. Not in the sense that he needs to know your personal details, and certainly not in the sense that he would accept bribes or operate in a corrupt way, but in an honest and old-fashioned sense—in the sense that it is necessary to know and trust and appreciate the person across the desk from you before making any agreements with them. Raj is a master of ceremonies, the person who everybody waits for and expects to speak to the group, to provide leadership, to communicate solution to the group, both intellectually and interpersonally. No professional event would start without words of wisdom and motivational encouragement from Raj. Before working in airline cargo operations, he was an aircraft engineer. He helped to design the LD-36 ULD cargo plane and belonged to industry associations such as the International Air Cargo Association and the Air Cargo Committee of the International Air Transport Association.

Since he was an airline expert, I leveraged the opportunity of being with Raj to continue my quest to update the D-Factor profile of an airline ticket. I asked him about air traffic on the New York—Cairo route. The 162 countries that signed the Paris Accord agreed nine years ago to tax airlines on their CO_2 emissions. Until 2016 airlines had been completely exempt from any emissions standards. Then there was a period in which some EU rules came into effect, but these did not affect flights to or from Egypt. Finally, there are the current rules, which affect all countries and all routes, passenger and cargo. To cover the cost of the new rules, the airlines have installed more fuel-efficient engines and purchased newer, larger and more fuel-efficient planes for long-haul routes, which include the New York—Cairo route. That reduced the cost of compliance per passenger by about 50 percent for those routes. The airlines also restructured their network so more long-haul routes with big planes connect in "hub" airports to shorter-haul routes that use smaller feeder planes to carry cargo and passengers to their final destinations. The optimized balance of long-haul trips and smaller cheaper-to-operate jets on the shorter routes cut CO_2 emissions, and also costs, by about 20 percent for the major airlines. The airlines then passed on the remaining 30 percent of the tax onto passengers in the form of higher ticket prices for those routes.

I ask Raj how operations in Cairo had changed over the last decade. He introduces me to Kareem Al Latif, the airline operations manager in Cairo. The two of them are about to have lunch, so the three of us walk to the banquet in the ballroom together.

Kareem is a tall, dark-skinned, brown-eyed Egyptian man. In Western dress, he looks Latin American. In a thobe and headdress, he fits the part of the well-educated, polite, wealthy Arabian Gulf businessman. He has four children, despite his relative youth. He has worked in Europe and Asia. Originally raised in Kuwait, Kareem is a sort of elite refugee. His family was driven out of Egypt when pro-el-Sisi forces overthrew Mohamed Morsi, and after living abroad ever since, he finally returned last year. Kareem is relatively shy. He comes from "the old school." He believes that words count, and as a gentleman he assumes that people listen. As long as those premises are true, he is a very effective manager. Kareem and I exchanged contact information, and I know we will take advantage of numerous opportunities to visit with each other in airports and cafés around the Middle East and Europe. I know I will enthusiastically follow his professional and personal development. I ask him if it's OK for my son to call him for help on a school project about Egypt, and he says yes.

As is customary with these conferences, we get on buses and take a tour of at least one facility to get a "hands-on" view of operations in the city where the conference is held. Regular conference-goers have seen many facilities over the years, often in dream locations. The conference organizers try to select the most desirable locations each year to attract the most attendees. While the most interesting sites from a technical point of view might be Karachi, Dhaka, and Damascus, the conferences may end up in Barcelona, Dubai, and Honolulu. Regular attendees write off their junkets to these exotic places on their corporate expense accounts. Cairo has come a long way to be included on the conference itinerary.

The Cairo airport is modern and organized, unlike the airport I experienced in 2010. Back then it was a free-for-all, no rules about anything. Airport queuing was a disaster, security was a half-hour exercise in futility (I could have gotten in with any contraband items despite the long lines and pat-downs). Now the operation is well laid-out, clean, and efficient. Technology like check-in at automated kiosks, scanners and Radio Frequency Identification (RFID) for immigration control, and mechanization including driverless trams, monorails, and customer service robots, are used everywhere, so the foot traffic is free-flowing. This translates to fewer low-skilled jobs in visible positions, but many more jobs behind the scenes, especially in IT.

I ask Kareem about how air cargo has changed on the New York to Cairo route over the last decade, since I came from New York the previous day. He said that the airlines now have to respect night travel restrictions to minimize noise in the nighttime in consideration of the residents living near the airport. At first they were concerned they would lose money due to the restrictions, but the nighttime flying restriction is now irrelevant. The city decided to build a new airport farther from the city where they could fly in and out without restricted hours. This turned out to be a win-win situation for everybody because it also reduced traffic congestion in the city and simplified passenger access to the airport.

There was also less crime in the port compared to ten years ago. After my last trip in 2010, I received a call from a wholesaler of apparel whose containers were stolen while being transshipped in Cairo. The thieves surely did not know that the containers were loaded with oversized maternity T-shirts which were practically useless to anybody who wasn't female and *very* pregnant and wanted everybody to know it from silly emblazoned cartoons on the front. It made me laugh and reminded me of another comic incident in which thieves stole whole containers of shoes. Too bad for them that the containers they stole were loaded only with shoes for the left foot!

Due to the new technologies, clearing customs in Cairo is effortless. There isn't even a line. An invisible reader scans and analyzes my chip-embedded passport and lets me through without delay (some other people get diverted to a line for additional security). The interior of the airport is clean and attractive, and is filled with sculptures and art. There is almost no queue outside for taxis, as most people use ride-sharing apps on their smartphones and there is a separate area for ride-sharing pick-ups. This is a far cry from the airport I arrived in when I came to Cairo in 2010, which had a few small utilitarian shops selling home appliances and a taxi area swarming with con artists where at least eight people approached me as I walked out of the airport, soliciting me for their taxi services, or the taxi services of a friend of theirs.

The ride from the airport to the city is also much more pleasant than before. Before, the air was highly polluted. On the street, buses billowed exhaust every time their drivers stepped on the gas. Exhaust, lacquer, paint, and other toxic fumes made their way into my hotel room on the twenty-third floor. Piles of trash littered the sidewalks. Now the air is cleaner, thanks to a UN-sponsored global air pollution standard for metropolitan areas that limits PM 2.5 to a level that is 50 percent lower than Cairo's level in 2010. The trash has been removed as part of human zoning principles formalized by the American Institute of Architects four years ago.

As I was coming toward the close of my trip in Cairo, I prepare my brief for Alex so he can crunch the D-Factor numbers for the airline ticket. Alex does some research and analysis, this time updating the D-Factor for a plane ticket, but this time from New York to Cairo instead of Detroit to Abuja since this is the data I have fresh at hand. Here is his analysis.

The Depletion impact of the ticket has dropped by 30 percent, says Alex. The United States instituted a carbon tax which started at 2 percent and is programmed to rise progressively to 18 percent, which inflated ticket prices by about 9 percent and required airlines to operate more efficient routes and upgrade to newer, more efficient aircraft for the long-haul routes and smaller, more efficient ones for the shorter feeder routes. Relocation of the Cairo airport to the exurbs (the far suburbs), better airport fume control, better integration of the Cairo airport with its community, and revitalization of the New York and Cairo airport interiors have driven the Devastation factor down by 43 percent Better transit links to the Cairo airport, and elimination of sexual discrimination among flight crew, has reduced the Dehumanization factor by 72 percent. Although there weren't that many changes in the supply chain for airplane tickets compared to more complex products like electronics and automobiles, the few changes that had occurred delivered very significant benefits. Overall, they reduced the average D-Factor for a transcontinental flight by 36 percent.

Mariam invites me to stay around for the sacrifice of the lambs the next day. She tells me that all the households are busy bringing in lambs today in preparation for the ceremonial sacrifice tomorrow. I tell her I appreciate the offer, but I will have to witness that on the next trip. I have a commitment to speak in Abu Dhabi tomorrow on the state of renewable energy and the rate at which various technologies would supplant the older, conventional ones. Ever since I completely missed a speaking engagement by mistake, I have been nervously conscientious about being on time for them. A few years ago, I put a Boston speaking engagement into my calendar while I was in Las Vegas, and it went into the calendar on Pacific time, which is three hours behind East Coast time. The day of the event, I sat quietly in a restaurant having breakfast while fifty people waited for a half hour for me to show up. I wandered in an hour later, thinking I was early. Never again!

FIVE

Abuja to Dammam

JUBAIL INDUSTRIAL CITY, SAUDI ARABIA, MAY 3, 2010

The Middle East is a paradox—on the one hand, the warmest, friendliest, most personable business interactions I have ever had, and on the other hand the most swindlers and fraud I have ever experienced. On my first assignment in the region, in North Africa, I remember my client walking arm in arm with me down the street. The intimacy would have been completely inappropriate in an American for West European business situation, but in this environment it was helpful in bonding and aligning the interests of the World Bank, which I was representing, and the government of Tunisia, in important restructuring. And, it was just plain comforting to know, or to believe, that there was a deep human connection and purpose behind the work that I was doing. In Saudi Arabia, businessmen kissed each other on both cheeks. There was a lot of touching, and it was very important to say hello, to stop what you are doing and socialize with workplace peers, without much regard to their status or level in the company. Some successful businessman spent so much time chatting and hugging, and with small talk, that it made a forty-hour workweek effectively thirty or thirty-five hours long, and may have been a more effective way of achieving what the Swedes attempted to get out of a thirty hour work week—working less and at the same time being more productive.

On the other hand, as if to offset "warm fuzzies" of the human connection, over my thirty years in global consulting all my experience with fraud and nonpayment seems to have come from clients in the Middle East, the Indian subcontinent, and Africa. I once formed a business partnership with the husband of a colleague of mine. Early in our conversations, Biwar Al Fakir mentioned that he had attended Yale University in the United States and graduate school at Oxford University in the United Kingdom. He was polished and impressive. He was tall, thin, and dark-skinned, and wore comfortable designer brands and custom-tailored Western clothes. He looked ready for a cocktail party or get-together at the golf club, and spoke eloquently and with a slight British accent. I only found out his roots by asking and digging; he did not volunteer any information. By origin, he was Pakistani, from Lahore. He did everything he could to mask his roots, including lying about his college and graduate school education. He was hesitant to provide his resumé, which seemed to be out of modesty and humility, in the way that graduates of Harvard University find ways of never saying the name Harvard because the name is so charged with elitism that people might perceive of them differently once they know that they graduated from there. When he finally provided his resumé, it confirmed that he had attended Yale University and Oxford University, but he politely declined when I said we were going to put his professional biography on the company website. "Not necessary," he said, as he sounded humble. "We don't want to make clients feel uncomfortable; not necessary to brag," etc. The pretense of having attended these two top schools give him open access to many job and career opportunities that would've been closed off to him, and gave him unspoken credibility in meetings and business situations. People gave him the benefit of the doubt, even if his knowledge was scant and his advice was unconventional. Unfortunately, his knowledge *was* scant and his advice was *wrong*, and he had never attended or graduated from either Oxford or Yale. Like Frank Abagnale, whose story was memorialized by Leonardo DiCaprio in "Catch Me If You Can," he was an expert con artist who covered up his ethnic roots and lack of quality education with a thick artificial façade of class and distinction.

I had experienced at least four other incidents of fraud and nonpayment in the region. In one case, a Saudi local agent collected money from a contract for my company, then failed to pass the money through to my company, having spent the money on his own account instead. In another incident, two managers at a United Arab Emirates oil company copied and pasted an entire article from a California State University professor, said it was their own, and then, to add insult to injury, asked me to say it was *mine*. The other incidents were nonpayment of relatively small amounts of money that had all been preagreed in writing, for example a construction firm in Qatar that had agreed to pay for my international travel expenses and then never paid the bill and did not respond to follow-up calls. In 2010 I was becoming increasingly familiar with this pattern.

In this case I was on my way to see Badr Al Ebrahimi in Dammam, but first I had some nonpayment business to take care of at a client in Jubail. The petrochemicals industry is directly linked to the oil and gas industry since so much of its input costs are for natural gas. Hassan Al Nopayit oversaw large new petrochemical plant projects. He didn't actually manage the projects; project managers were in place for that. He oversaw the project managers' management, which means he got in the way of things getting done. He was a short man with a tall ego who wore a Cheshire smile that telegraphed a false sense of congeniality. The more he smiled, the more I knew something was fishy. Despite the fact that our project had been completed, he delayed payment for a year, playing an extended, slow-motion game of cat-and-mouse. Sending an invoice, which is the only step needed in the West, was only the initial trigger in a dense morass of procedural impediments to getting paid. For generations, foreign companies doing business in Saudi Arabia have needed local Saudi agents to help them advocate for each payment that is rightfully due, to counterbalance obfuscatory walls of bureaucracy like this.

After I left Jubail, I traveled along the highway to Dammam, Saudi Arabia's second largest city. Traffic slowed down in one spot, and I saw why: a car had sped into the back of a trailer and there was not much left of the car. It looked like it had just been through one of the crusher machines that junkyards use to compress vehicles into small cubes for disposal. From the looks of the scene, I'd say the car must have been going about 100 mph and the trailer had probably been completely stopped in the middle of the road without any lights on. I also saw a tractor-trailer jump over a three-foot high concrete barrier in heavy traffic. Saudi Arabia has the worst vehicle accident rate of any country in the world. I suppose that crazy driving comes at least partly from the repressive regime, which causes people to have many hidden vices, including drugs, that can't be seen except in the autopsy after you kill yourself, and partly from the low level of education of many of the migrant workers like truck drivers, many of whom have no professional training and minimal education. Traffic deaths may also be related to an ingrained habit of dodging laws because the laws are often not practical, so drivers use the breakdown lanes, weave between lanes, and anything else they feel like, and because of a general lack of law enforcement, which includes traffic laws.

As I left Jubail, I saw the huge and continuous gas flare, which was visible from far-away towns. Trash and rubble were piled in front of many buildings. Burnt chemicals hung in the air and became baked into dust from the sandstorms. "Canteens" and convenience stores provided ultracheap food; there were few restaurants and they serve almost exclusively fatty and greasy fast-food. Jubail was an industrial city. Over half of the inhabitants were laborers from Pakistan and India on temporary work visas. Women were not allowed in the industrial areas, and they were rarely spotted on the residential streets. There were no walkways to cross the broad avenues, and vehicles seemed to have the right of way at all times. Cars and trucks were often in ill-repair and spewed exhaust. Although there were detailed laws, regulations, and standards governing everything from environmental health and safety practices to driving and workers' rights, a void of education, training, and enforcement somehow allowed a large degree of work, life, and business to occur outside these boundaries.

I made my way to Dhahran, where my company was bidding for a large contract with ArabaCo. Saudi Arabia was the largest oil producing country in the world. In 2010 it accounted for more than 20 percent of proven oil reserves and more than 17 percent of the world's oil production, and that was down from the peak. In 1976, when I was a child, Saudi Arabia produced 37 percent of the world's oil.[1] Saudi Arabia shares extreme oil and gas wealth with other countries in the Gulf Cooperation Council (GCC), the group of six Arabian Gulf countries that also includes Qatar, Kuwait, Oman, Bahrain, and the United Arab Emirates. Qatar was the largest producer of natural gas.

Badr was our agent in Saudi Arabia, and I needed to discuss the contract with him. Badr was a tall and statuesque Saudi Arabian man with sculpted features and black hair that was both wavy and curly. He was handsome, almost the Arabic equivalent of Michelangelo's statue of David. He had spent twenty-five years at Saudi ArabaCo, including in the overseas branches, before retiring the previous year. After retiring, he helped develop ambitious Kingdom projects such as the King Abdullah University for Science and Technology.

At sixty years old, he had a grade school girl at home, a high-schooler, and three adult children—offspring from two ex-wives and two current wives. In Saudi Arabia, I had to get used to the fact that when men spoke about their "wife" it could have been any one of multiple women, and when nondivorced men referred to their "first wife" they may have meant the woman they married first, or their ex-wife, which could be different women. The distinction was not just semantic: despite the fact that men were allowed to marry up to four women under Islamic law, the divorce rate in the Kingdom was 30 percent. First wives, second wives, and ex-wives were all very common terms in everyday conversation.

Badr spoke with confidence and passion, and was always ready to speak in a formal, old-fashioned sort of way on a wide repertoire of subjects that he holds dear. In the Saudi tradition of thinking big, his ideas and his actions often went "over-the-top." Badr invited me to one of his houses, and laid out a feast fit for a king. I did not have a chance of finishing 10 percent of what was on the table, but as the host he graciously insists on stuffing me like a turkey at Thanksgiving. Over the ensuring years we would share many memorable moments during car rides, dinners, and business meetings, and personal conversations.

At one point in our discussion he politely excused himself to pray. All executives in Saudi Arabia stopped to pray. Meetings stopped, and stores were closed during prayer. Quietly (slight call from the mosques) the store doors all closed. During this time there was no shopping, no commerce, no food, nothing happened during this time, and it happened five times a day, usually for thirty to forty minutes apiece. The ubiquitous presence of prayer in daily life in Saudi Arabia was remarkable to me as a Westerner and non-Muslim. The TV in the lobby of the hotel was continuously and permanently showing throngs of people in Mecca around the Kaaba. Even the TV in the taxi stand (a public facility, I guess) was tuned to men prostrate in the mosques.

On the way to the hotel, my driver got into a heated exchange with another driver. I didn't understand the Arabic language, but the hand and body language was clear enough for me to understand what was going on. To enter the hotel property, we went through a security inspection. The guard examined the trunk and underside of the car. This happened every time we entered a compound of any kind. Two camouflaged jeeps hidden in green branches were sitting outside the front of the hotel with machine guns, ready to respond in case of an attack.

DISCOVERING UPSTREAM OIL

In order to produce so much oil, the Kingdom needed a lot of equipment and oilfield services. So ArabaCo bought a huge amount of equipment, services and supplies, in fact more than just about any company, period. My company was making a business proposal to ArabaCo.

To get into the building where our meeting was to take place, we had to first pass barricades on the street. Then there was a military guard looking attentively at everything and everybody, with his machine gun poised on one of the barricades and his finger on the trigger. I walked directly in front of it. I had never walked directly in front of a machine gun pointed at me before with somebody's hand on the trigger. Once inside the building, we underwent three layers of low-tech physical security (pat-downs, metal detector wands). First, they tried to find our name on the list at the outside security gate, and it wasn't there. The security guy flipped through every page of security permissions in a big book, at least twice, while the three of us peered down to see if we could find ourselves. Big security breach right there, because we could see all about everybody who was seeing everybody else and why. Second, three outside pat-downs. There were always a gaggle of young men in uniforms hanging around, and there was always one who stopped you authoritatively with a barrier of some kind (machine gun, wooden stick, something physically intrusive and preventive). They didn't care who you were or when your appointment was. Third, the inside security: X-ray machine for baggage and the same pat-down and check of who you are going to meet. It was precisely redundant with the ones outside the building. I received less scrutiny than the two non-white people behind me.

The point person for the meeting was Lou Lewis, a procurement manager at ArabaCo. Lou was an imposing British mature businessman. He carried a slight weight around his whole body, which lent gravitas and credibility to what he said. He had gray blonde hair that had slightly thinned over the years, but which kept him looking young and fashionable despite his sixty-three years of age. He spoke emphatically and enthusiastically out of intellectual curiosity and drive. He had an inner motivation that came from someplace deep in his gut. It was always there.

Lou always repeated back what you said, so as to confirm and "handshake" with you, which established a rapport and trust. Even when the conversation was going rapidly, he repeated the first few words of what you said before launching into his own response. With every sentence, he continued to build rapport and made sure you knew he had heard you. Some people do this occasionally at cocktail parties, but Lou did it in every conversation, even if he had known you for thirty years. While some people considered it annoying, it made him easy to access and be with.

A Brit by origin, Lou lived as an expatriate in Saudi Arabia for his entire professional career. The Kingdom had truly become his home. As he immersed himself in his work, and enthusiastically and passionately pursued his daytime job and his off-hours teaching, studying, and quasi-professional goals, he never felt constrained by the limitations of living in Saudi Arabia.

There were forty-five people in the room, all dressed in headdresses and robes. The senior vice president gave a formal introduction. Everybody deferred to him and they made sure I knew he was entering the room when he came in so I could focus my attention on the Big Cheese. My presentation went exactly as planned for sixty minutes, and then we took a five-minute break and reconvened and they asked questions for an hour. The Big Cheese spent an hour discussing what he wanted to see a little differently here and there. At the end he went around the room and made sure every one of the forty-five people on his team also gave individual feedback on our report and my presentation, even though he dismissed and overruled about a third of them.

There wasn't much to do in Al Khobar. You could get a Rolex and digital camera and a BMW or Lexus, but it wasn't a great place to enjoy any of those things. Outside of the shopping areas a lot of the land consisted of sand, dirt, and rocks. I was with Tom Brown, who was serving as a consultant to me. Tom had recently retired from an executive level position at a multinational chemical company, where he was responsible for operations. He had controlled billions of dollars of expenditure and spent six years as an expat in its French operations. He had an even-keeled temperament and knew how to read the people in a meeting, which was useful in our meeting with so many people.

Tom and I finished at ArabaCo at noon and our flight didn't leave until 1:25 AM (13.5 hours later). The shops, and everything else, in Dhahran and Al Khobar were closed for prayer from 12:15 until 3:30, so we opted to spend our free time the next afternoon in Bahrain, where shops and restaurants didn't all close during prayer. Bahrain even allowed Western pop stars to be broadcast on TV, and hotels served alcohol. In fact, there were so many hotels that I think some hotels may have existed just for the liquor license.

In downtown Manama, buses were belching exhaust and throngs of people from India and Pakistan mulled around. There were many construction sites, but few construction plots were near completion. It looked like they were started a few years ago and left unfinished, resulting in a whole lot of dirt and bricks lying around. Barbed wire, don't go there, like East Berlin before the wall fell. We tried to take a few pictures and were told not to. I thought nobody was around, then a guard shouted "Excuse me!" Very polite, with a smile. Tom and I stopped for coffee and sat near two people who smoked shisha. We walked to the water's edge and found a tour boat, but there were no passengers. The operation had been suspended after fifty-five people drowned two weeks earlier because the boat was overloaded. They would be reopening again the next week, with some tighter regulations.

Women in both Bahrain and Saudi Arabia were for the most part fully covered. When I returned to the West I noticed how outrageous the Calvin Klein ads and girls' scanty, tight clothes, boob jobs, and bare midriffs must have seemed to the Saudis. Men's dress in Saudi Arabia and Bahrain was simple and clean—full white robe, and white headdress with a black ring around the top to hold it on. The main variation was the headdress, which varied in style by nationality, and was either white or white with red checkers.

"Once the oil is produced, how does it get to market?" I asked Badr. Badr drove me in his Chevy Tahoe to meet with someone named Muammar. Driving places with Badr was a learning experience. As he drove, he talked about politics and religion and culture. He was passionate, but ready to joke or appreciate some light humor, about Islam, and religions in general, about Iran and the network of terror it spread throughout the Middle East (seen through the eyes of a Saudi, of course), about American political interests in Yemen, Egypt, Israel, Iraq, Syria, and of course Saudi Arabia.

Muammar Khan was a dapper sixty-three-year-old Lebanese man whose face, body, and demeanor could all pass for thirty-seven. He wore fitted, designer suits and thin ties on tasteful white shirts with pinstripes, which helped project an aura of youthfulness. He smiled at the right times—whenever he should be smiling, he is, and it seems genuine. Photo opportunity? He was ready in a heartbeat. All of this made him the consummate salesman. Muammar used a motto—"never give up"—on all his emails and Skype conversations. Indeed, he never gave up—he was doggedly persistent. This later served him well in his role as my company's Country Manager for Saudi Arabia. He was an Engineering graduate and a project manager who had graduated from the University of Petroleum and Minerals (KFUPM) in Saudi Arabia.

Muammar explained that drilling can yield oil, gas, or both. Most holes yielded a combination of oil and gas. The oil supply chain began with drilling, often with the help of a drilling rig, which could be either onshore or combined with a kind of floating vessel for offshore drilling. In the oil industry, this phase was called "exploration." When the drilling yields oil and/or gas, the well went into "production." Once the oil that was produced came up out of the ground, it was usually carried away in a shuttle tanker or very large crude carrier (VLCC), if offshore, or transported via a pipeline, if onshore, to a marine terminal, and onward via rail, truck, or pipeline to a refinery. The natural gas supply chain was a bit different. Gas was preprocessed as it came out of the ground, to separate liquid gas from gaseous gas, which made it easier to transport to the different facilities that process heavy or light, wet and dry, gas. This was called natural gas liquids (NGL) fractionation. Preprocessing was done by midstream operating companies, and the goal was to end up with pure methane gas, by separating out the ethane, butane, and propane.

It was getting late, so Muammar and I went to dinner. As we drove, I started noticing the *absence* of things. For example, there was no music. The breakfast room in the hotel was always a silent place (I had heard that it was illegal in Dubai, which was also in the GCC, to play music in public). Second, there was very little noise in public. In the Dammam airport, I passed a group of young American ladies who were greeting each other happily and noisily, and a security guard came over to shush them. People were organized and quiet in the way they did things. At first I thought, oh that's nice, but then I connected some dots to repression and intolerance for dissension. What would happen if dissidents were loud, or disobedient, or said the wrong thing. The country had about twenty-seven million people, of which I heard that well over eleven million were immigrants, mostly Pakistanis, Bangladeshis, Indians, and Philippinos. All of those people were here on a work visa at the grace of the Kingdom's legal bureaucracy, and were promptly deported if there was any irregularity in their legal status. The stakes were too high for them to risk making waves over anything like working conditions, benefits, or freedom of speech, even if they were abusive. One Saudi colleague referred to this in passing as "the slave thing."

Third, women were generally not seen much in public, and when they were in public they were fully covered. I really didn't see women on this trip at all until I went to a trade show in Bahrain. Women just weren't walking around in public much in Saudi Arabia. They were required to be accompanied by a husband or other male family member when in public, and they couldn't drive, so this greatly restricted their movement. The separation of women was further formalized by the fact that singles and families were separated everywhere, and the definition of singles excluded women, so women could only dine in the family section, and to dine there they needed family with them, so basically they could not eat at restaurants without their husband, and if they didn't have a husband they needed to get one. It was not obvious at first. You think—oh, they have a separate entrance labeled "Families." That's nice. Then you realize after awhile that it was like this at every single establishment of any kind. Even in the hotel restaurant, there were two rooms and two separate entrances. I thought it was just two different rooms to sit in, but it was a form of mandatory segregation.

Since I was becoming so knowledgeable about the oil and gas supply chain, I decided to collect just a little bit more information so I could send Alex enough data to get a D-Factor on a tank of fuel. I was curious to compare that to the D-Factors for a cup of coffee (1.8) and a transatlantic flight (1,385). I paid a visit to someone I was told knew more about oil tanker operations than anyone in the world: George Kendrick, the General Manager of Osiris, an oil tanker company operating in the Arabian Gulf. George had an office in Dhahran, so we met that same afternoon.

George Kendrick was an extremely fit man, with a tight build and no fat, evidence of a career in the military and a lifetime habit of push-ups, setups, and crunches. He was of normal height, but had a focus in his eye that let you know he was more important than you and he was judging you. He was American, the oldest of three children from the Midwest. His father was military, too, and the family located in a number of rough spots throughout his childhood—Afghanistan, the Philippines, and Sudan. This prepared Tom well for managing a vast oil tanker fleet that transited the Strait of Hormuz daily and encountered armed pirates from Somalia and Ethiopia. In high school he played four varsity sports and captained the cross-country, rowing, and wrestling teams. Tom held his cards close to his vest, keeping a poker face in all circumstances. As a father myself, I wondered if his aloof mannerism and stance made him a difficult father. After Tom had graduated from the State University of New York with a degree in oceanography, he took a commission as ensign in the US Naval Reserve and got his Coast Guard third-mate's license. After three years at sea, he worked at the biggest American oil and gas company for over twenty years, including stints in London, Monte Carlo, and the Louisiana coast, before running crude oil shipping operations from Dubai.

Osiris has originally been a crewing company. Then it bought product tankers and became an end-to-end tank shipowner, operator, and charterer. It shipped oil from countries in the Arabian Gulf to refineries in the United States, particularly serving the coasts of Saudi Arabia, which had large petroleum ports at Jeddah, Yanbu, Jubail, and Ras Tanura. Although Osiris had a few dozen ships, I was shocked to learn that its fleet was relatively small compared to the world fleet. There were thirteen thousand tankers in the world fleet, of which over eleven thousand were oil and chemical tankers and the rest were liquid gas tankers hauling both liquified petroleum gas (LPG) and liquified natural gas (LNG).[2] Of the oil tankers on order in 2015, fifty-one were VLCCs (supertankers capable of transporting two million barrels of oil), forty were SuezMax (capable of transiting the Suez Canal fully loaded), and seventy-two were Aframax (medium-sized) vessels.[3] The tankers went to liquid terminal ports, where terminal operators offloaded the crude to refineries.

The company had a lot of its operations in Dubai. Dubai was building itself up to be a major vessel service company, including financing, crew support, and maintenance and repair, so Osiris's office there had easy access to crews (visas, immigration, and security were not as tight as in the other countries), and there were many ship maintenance shops along its coast for convenient lube, crew changes, and spare parts.

There were many pirate attacks during this period, including ships like Osiris's that were transiting the Gulf of Aden. The movie "Captain Phillips" with Tom Hanks memorialized (later, in 2013) the experience of one pirate attack on a container ship, and George Kendrick lived through the experience more than once in his role as the general manager of Osiris.

I understood the shipping part, but the terminal part was a bit fuzzy, so Tom referred me back to ArabaCo to talk to a logistics manager named Darren O'Reilly. When I arrived in Dammam airport, the driver, Akbar, picked me up to drive me to the importer at ArabaCo. Akbar was a tall, lanky young Saudi man, about twenty-seven years old. He had short, black, greased-back hair that was youthfully stylish, and he wore casual Western clothes—chinos and a sort of Hawaiian shirt, which I think was a deliberate statement of rebellion from the customary Arabic expectations of formal dress and mannerisms. He spoke no English, so took direction by sign language and Google Translate if there was Internet. Following the custom of the land, he usually wore a four-day beard in honor of the prophet Mohammed.

Although every Saudi was religious—it was not tolerated, barely conceivable, in the minds of Saudi Muslims, to be agnostic or atheist—Akbar did not stop to pray during the day like the others. Maybe he snuck it in late and while he was alone (that is allowed to a degree, but you get less points), but if he did, nobody saw it.

Akbar drove me and the team members wherever we needed to go in the Eastern Province, which usually meant Al Khobar, Dammam, Dhahran, and to and from Bahrain. He drove through the sand dunes and dirt roads in an old Lincoln town car, which sounds luxurious but was an unpleasant safety hazard. The seatbelts had been removed, perhaps to stop the dinging noise on the dashboard when nobody occupies the seat.

The engine was in ill repair. It had probably never been tuned up or maintained since the car was new. The valves tapped every time he accelerated, the exhaust smelled like unburned fuel from inside the car. Even with the windows closed and the air conditioning on, the fumes of burned and unburned gasoline were suffocating.

Akbar gave no outward signs of being unhappy, but his eyes were glassy some days, and on those days he was untalkative and drove like a bat out of hell, tailgating at 90 mph on the Causeway (the 25 km bridge that connects Bahrain and Saudi Arabia) and spinning wheels on fast starts and around turns. He was clearly taking his inner angst out on the car, and by extension, its passengers. I asked him on several occasions to drive slower and more carefully, but between the language barrier and a desire not to hear the message I was conveying, it did no good. At one point, I almost asked him to drop me off on the side of the road so I could find an alternate ride. This would have come at a serious cost to me, as there were no other cars around and with no charge left on my Saudi cell phone, I would have been left essentially stranded in the hot desert with no alternative means of transport, but better that than dead.

Even without Akbar's unique blend of passive-aggressive rage and automotive hostility, driving in Saudi Arabia was dangerous and annoying, and getting across the Bahrain-Saudi border was an adventure. As you approached the border, the road became severely rutted and the speed bumps were so aggressive they cracked your spine. Perhaps this was all intended to "break" you before you faced the customs officials! During alternating trips, I tried to bring hot coffee in the car, read messages on my phone, and sleep, but none were possible. Going from Bahrain into Saudi Arabia, there were seven checkpoints, each of which required its own queue. The first was for buying insurance for driving in Saudi Arabia, if you didn't have it already. The second was to check your car insurance papers, which you may have just bought at the immediately preceding checkpoint. The third was to stamp your passport upon exit from Bahrain. This one took some time because they checked your entry dates and exit dates, and how much you paid for your entry visa, and determined if you owed more money due to overstaying in Bahrain. The fourth was to check your passport upon entry to Saudi Arabia. This one could also take time because the visa requirements for Saudi Arabia are strict—you needed an invitation, and getting all the paperwork in order could take months and often resulted in rejection—and border agents checked all the details. Small discrepancies could result in rejection or sidelining to the customs buildings off to the side, which was like getting put in the corner as a kid for punishment. The fifth was to check your car insurance papers in Saudi Arabia. This one should have been an easy "pass-through" because it was just a matter of collecting a piece of paper, but at this gate they corralled all the cars into one lane, which created a severe bottleneck. The sixth was for security officials to inspect the vehicle. Here, they checked the trunk and the inside of the vehicle and, I am told, sometimes felt between the seat cushions for bootlegged liquor. On Thursday afternoons (Thursday is the last workday of the week in Saudi Arabia and Bahrain), the formalities of crossing the border from Saudi Arabia to Bahrain airport took two to three hours.

I once asked a driver (a taxi driver, not Akbar) why they constrained the fifth step to a single lane. Why don't they just open up all the lanes and keep the traffic flowing, or better yet, use electronic transponders and get rid of ALL the checkpoints? There was no good answer. It seems that the culture of bureaucracy and control developed to such an extent that people looked at the chaos as a way of life, even fun, without which life would have been boring. Some people got angry or annoyed, but most turned it into a social event, joking and socializing with the people in the cars next to them and with the customs officials. Some people did both, cutting across the lanes and disrupting traffic while joking. I looked at it as a ritual that should be studied by cultural anthropologists.

I went through the usual security layers to get into Darren's office. Like George Kendrick, Darren was firmly built. He had also previously been in the Navy. He ended up in midlevel management of day-to-day operations, not senior management or the executive ranks. He was a friendly, affable guy of moderate competence—the kind of person who had a lot of friends and got things done because he had many drinking buddies. Physically, he used to play football, but it looked to me like he had not maintained the same physical fitness regime as George Kendrick.

Darren's role didn't actually have a job description. There was a requisition number in the SAP system for his post, and there was a box on the organization chart, but nobody actually defined what he was supposed to do. Most of the time, the tank truck drivers and the storage tank operators and the repair technicians moved the liquid product from the port inland to the refinery without his assistance, and they preferred that way. They knew what to do, and any interference from management was an unwelcome disturbance. Besides, management and operations were like different animal kingdoms, and letting animals from one habitat have encounters with animals from the other tended to cause problems.

For inbound tanker shipments that came to ArabaCo's refineries, Darren was responsible for bringing the vessel into port, making sure it got unloaded, and sending it off again. Then managing the physical storage of crude as it entered the refinery. It was a pretty simple operation, as logistics goes. Some product was transported by pipeline, and for those shipments he needed to also schedule batches to be released through the pipeline at the right time to fill the pipeline and not run into capacity shortages.

Since most of the world oil was shipped from the Middle East, the Arabian Gulf was typically jam-packed with oil tankers—at any time, fifty or more tankers were located inside the Strait of Hormuz.[4] The liquid product got loaded into a tank farm, which consisted of a plot of land adjacent to the ships berths, with a lot of piping between the loading area and various types of tanks. There were LNG tanks, crude oil tanks, syncrude tanks, diluent tanks, gas storage tanks, and diesel fuel tanks. They all looked similar to me except the spherical natural gas tanks, but they held different products and were designed differently to accommodate each product. A loading arm connected the vessel to the tank, and various other equipment helps to offload the fuel. The terminal was full of electrical circuitry, pumps, pressure vessels, mixers, and control systems.

Getting product into and out of the Kingdom was a hassle, and Darren was often called on to work through the dense bureaucratic fog. Product was usually detained in Customs for days or weeks, requiring a wad of arcane bureaucratic paperwork and stamps, which had to be complete, in order to be released. The process was so inefficient, cumbersome, and lengthy that companies would ship to and from Bahrain, then transport the product from Bahrain into Saudi Arabia, which was easier.

Getting people into and out of the Kingdom was even more excruciating than moving products through customs. There were quotas by job title (engineer, technician, etc.), nationality (Indian, Philippino, etc.), level of management (manager, supervisor, etc.), and type of company (trading, manufacturing, etc.), and the quotas were additive; for example, a Saudi manufacturing company could only have a certain number of "Pakistani" "supervisors" with a given "level of education." Visa applications were frequently rejected on technicalities. In addition, the country's Saudization laws have made it even harder to get foreign workers into the country for any kind of job.

It turns out that Darren didn't do much work. He occupied a loaded midlevel management position in a big, bureaucratic Middle Eastern organization where being a Western expatriate earned you a sort of lifetime pass. Sure, once in a while there was a presentation or something that needed to be figured out, but he found that copying and reusing material from previous presentations and from the web generally satisfied the needs of his superiors.

I dictated what I had learned, in shorthand, and emailed the voice recording to Alex so he could develop a D-Factor for a European consumer who fills his car with a tank of fuel that was supplied through this supply chain. After he listened to my voice memo, he said he could not give me the complete, Upstream and Downstream, analysis since I hadn't given him enough information on the refining and delivery to the filling station. But he came back to me with the Upstream portion. At the top of the page he had scrawled a handwritten number and circled it, like a teacher grading a paper. It said "Total D-Factor: 85." That was the equivalent of forty-seven cups of coffee. Then he added some of his observations on the Upstream value chain.

He said that the refineries I discussed in Saudi Arabia were so large that their dedicated power plants on the same scale as a municipal power station generated heat, steam, and CO_2 in proportions that made them the largest contributors to global warming of any industrial or commercial activity. Depletion factor: 62. Then he had some notes, which were slightly illegible, about the exhaust and air quality in Bahrain. Devastation factor: 16. He noted that all-male engineers determined the drilling plan, and that many were from Pakistan, Bangladesh, the Philippines, and Egypt. He further noted that they were imported under national agreements between the countries that have features such as repatriation once in three years, and that their social media conversations were monitored by the government, and political dissent was not tolerated. A bullet-pointed list contained items such as repression of women, including singles segregation, intolerance of photos (which he noted as "transparency"). Dehumanization factor: 7.

He noted a few questions about the lack of music and the unacceptability of joyful noise, but didn't factor them into any analysis He noted "PIRACY" in all caps, and "bureaucratic fog" in a sidebar, but also did not quantify those. "That's all I can say about Upstream," he said in his reply email. "Let me know when you have more information on the Downstream picture."

The physical buildings of the Dammam airport were glamorous—rich and new, marble—but service at the counters was strictly "third world." They were overcrowded with Pakistanis, Bengalis, and other migrant workers, and they didn't have the people, talent, or systems to match the infrastructure. In Dammam, Tom and I pushed our way through crowds to get our tickets after a solid hour in queues between 2:30 and 4:00 in the morning.

As a Westerner, there was a silver lining to this cloud. As a white American, I could jump ahead of the worst of the lines if I wanted. In some cases, the airport staff saw my white face and signaled me to move ahead to a different, shorter line. The other line was dedicated to Arabian Gulf country nationals or VIPs of some kind, but because I was a white businessman I got waved ahead. Even fellow travelers nudged me ahead in some cases, apparently aware of the norms. Once you knew this, you could take advantage of it, for example using body language to say that you expect to skip the line, or get other special treatment. "Act like a VIP and you will be treated like a VIP." It didn't work all the time, though. On one trip through the Dammam airport, a customs official felt that I handed him my passport discourteously, so he made me wait in the middle of a large hall until ten minutes later three other customs officials took me into an office and gave me a lesson on being American in Saudi Arabia. "You think you own us because you are American?" "You think you are better than us?" They made me apologize for nearly a half hour.

I'm not sure to what extent I did anything wrong, but after that I was careful to be nice to the customs agents.

Before heading back to Boston, I spoke at the PowerGen conference in Doha (Qatar) and visited the industrial city of Ras Laffan. The trip to Ras Laffan seemed like a much cleaner, more modern, and people-less Jubail. We drove for miles and miles past a landscape of huge pipes and cylinders, connected occasionally to vertical installations of processing equipment. They arose out of the desert sand, and are the only thing visible on the horizon. It was not a plant or an installation or even a complex. It was a "city," my guide clarified. A city of pipes, not people. The city of pipes had been built in phases by consortia of national gas companies, engineering procurement and construction contractors, and global equipment suppliers. This type of installation was at the top of the list of so-called mega-projects, which generally means projects ranging from about $2 to $25 billion in capital investment. For years, pipes were delivered and connected, instruments were calibrated, and terminals were built. The only people around were "industrial tourists" like myself and a handful of managers in small conference rooms. All transportation was by bus or utility vehicle; nothing can be walked. Ras Laffan transformed Qatar into a massive gas processing plant serving the entire world.

NOTES

1. Sarah Yizraeli, "How Important Is Saudi Oil?" *Middle East Quarterly*, March 2000, pp. 57–64. http://www.meforum.org/42/how-important-is-saudi-oil
2. "Shipping Statistics and Market Review 2016: Analytical Focus." *Institute of Shipping Economics and Logistics (ISL)*, Volum 60, No. 3. Bremen, Germany. https://shop.isl.org/
3. Source: Crude Oil Tanker Market Outlook, prepared by Affinity for Intertanko. November 16, 2016. Accessed September 28, 2017.
4. https://www.lloydslistintelligence.com/llint/tankers/index.htm

SIX

Cairo to Abu Dhabi

CAIRO, MAY 1, 2030

Cairo has become a model city with echoes of Paris and old Beirut, and reflections of the modern Middle East from Dubai, Abu Dhabi, and Masdar City. In fact, it has become such an iconic "new city" that I have been hoping for some business there as an excuse to do some personal tourism, rather than arranging some personal sightseeing around my business meetings in more mundane places.

A new layer of infrastructure has been laid over the previous city, preserving the old and providing a platform for rejuvenation and growth. The monorail from the airport provides a bird's-eye view of the clean and new infrastructure and buildings everywhere. The trunk line of the monorail runs down the center of long and wide Sheikh Hamed Road, which resembles the Champs-Élysées in its grandeur and the central role that it occupies in defining the city and directing traffic flow to and from the center. Very nice and centrally located public beaches attract international tourists including bodybuilders and celebrities, rivaling Jumeirah Beach in Dubai.

In contrast to the city I saw in 2010, Cairo today is clean, new, fresh, and efficient. The country's urban development plan for Cairo envisioned a low-carbon city, and the state of energy storage technology across a wide number of domains, especially architecture, mass transit, automobiles and buses, allowed that vision to flourish. It has become a center for renewable technology that puts the proverbially delayed Masdar City project in the United Arab Emirates to shame, perhaps because of Cairo's pragmatism in targeting realistic energy efficiency targets. In any case, the city's air pollution levels were among the best in the region. PM 2.5 is 35, about at the level of Newfoundland in 2017.

The city is cosmopolitan, partly due to the marketing of the excellent intuitiveness educational institutions as an export product that attracts the best talent from the Middle East, Africa, and near Asia. Due to its openness and progressiveness in social equality and human rights, it has become a magnet for foreign students, that succeeds in realizing the vision that was originally suggested by Saudi Arabia's King Abdullah University for Science and Technology (KAUST) near Jeddah, which attracted less than a thousand students per year,[1] mainly by luring world-class faculty with lavish packages. Many or even most of the KAUST students left the country upon finishing their studies.[2] For as grand as the vision was, women were not granted equal rights, and outspoken academics were imprisoned. In contrast, Cairo has followed a liberal, open path for its universities. The open and tolerant culture has spread throughout the city's life, which has resulted in a large influx of nationalities and religions, making it a "melting pot" on the scale of Dubai.

Cairo has also become a thriving business hub and financial center, based on civilized, trustworthy banking and contracting, Western and Indian institutions and principles. These principles are mirrored in the city's architecture, which has evolved to encourage human zoning including attractive residential and commercial clusters. The city is designed for walking and interacting, with a relaxing corniche that houses a concert hall which attracts world-scale concerts and athletic events. There is a massive new convention center, one of the largest, in the same category as Las Vegas and Orlando.

I am here to give a speech on the state of renewable energy and the rate at which various technologies would supplant the older, conventional ones. Cairo has also become the destination for the premier annual conference on renewable energy commercialization.

A wide array of regulatory changes has occurred since the last visit, which substantially changed the business environment for oil and gas, and consequently the supply chain of delivered fuels to power plants, industrial operations, and retail filling stations. First, Egypt's alignment with the UN directive on carbon taxation has resulted in an 18 percent tax on fossil fuels and a minimum of 50 percent ethanol-based gasoline, which of course was produced from renewable natural resources. Second, Egypt has transitioned from mostly natural gas-fired power plants to the point where nuclear plants supply over 40 percent of the energy for power generation.

Energized by the massive influx of foreign nationals and young students, the city has become a showcase for recycling. The universities recycle 100 percent of packaging for food and consumer goods. In addition, the country has coordinated with international lenders including the World Bank to develop a mandate for energy-efficient motors and appliances, as well as 60 percent or higher recyclable durable goods which includes automobiles.

The country's public endorsement of the principles of the UN Global Compact put pressure on businesses to improve social equality and human rights, and to report their performance each year in their annual reports. Many restaurants and stores adopted the D-Factor label, which business visitors and tourists have written about extensively, which has added to Cairo's image as a progressive modern city. In fact, the D-Factor label has become more prevalent in Cairo than government-mandated nutrition labels are in the United States.

REDISCOVERING UPSTREAM OIL

The conference hall is like the Las Vegas or the Abu Dhabi convention centers, where I have given my technical oil and gas talks in years past at the Abu Dhabi International Petroleum Exhibition and Conference. All the technical equipment and services work as planned. Each of about fifteen presenters gives twenty-minute presentations back-to-back and we keep to schedule. The hall itself is impressive in its scale, and provides an inspiring venue for the eighty thousand professionals who are congregating here to chart the future of energy technologies. The critical mass is approaching the levels seen at the height of the oil and gas boom in 2008 at Houston's Offshore Technology Conference.

Being a speaker at the renewable energy conference gives me outstanding access to other presenters, so I look up the speakers from the Egyptian national oil company. The CEO of the shipping division and the manager of production are both on the roster, and I try reaching them to continue my research into the oil and gas supply chain.

When I have a number of people to interview at a company, I usually start with less senior people so I can have a more informed discussion when I get up the C-levels. So I try "cold-calling" the production manager, Sameer Al Produsit, via email, since his email address was listed in the speaker directory. Sameer is in charge of production at the national oil company. His group purchased the inbound crude oil along with the materials and services that were necessary to process it, and I wanted to know how his job has changed over the last decade. Sameer is a shortish man with a pocked face who wears simple Western clothes—not tightly woven Egyptian cotton, first-class garments or by the executive class, but the kind you would buy at Target. Sameer is a nice guy, but outside of formal meetings, he only talks to other Egyptians. To outside vendors, including supply chain partners, this means that you still need an agent or middleman to be effective. However, Sameer is one year away from retirement, and he readily acknowledges that he is the last of his kind. Given the carbon tax and the momentum that Cairo and all its industrial infrastructure gained, his position will not be replaced. The new organization has diversified into a range of renewable energy technologies, and the company will likely be merged with the state-owned utility in a transition period before eventually being completely dissolved. Luckily for him, Sameer has been with the oil company most of his career and will have a very comfortable pension.

Since my topic at the conference was the transition from fossil fuels to renewable energies, and since a lot of my client work involved the oil and gas industry, my work already required me to gather information on the supply chain and operations, along with information on the ecological, environmental, and social impact that the industry was having. But by this time, I had taken on the additional challenge of updating the D-Factors from my travels two decades ago as a regular, day-to-day sideline to my work. Adding a meeting here or there was easy to do, and it allowed me to plug the holes in the database and come up with a more complete set of data.

For example, while in Egypt I also planned on meeting with the local agent and distributor for the bulk oil and gas—the equivalent person to Muammar. However, Sameer informed me that this person had recently left his post and could not meet with me. Given the stagnation in trading of bulk oil contracts, oil has become even more commoditized than in the past, and margins in trading business shrunk to the point where it was hard to make money in this business. The individual decided to switch from trading bilateral oil and gas contracts (swaps) in Egypt to trading natural gas future contracts on global markets on the London mercantile exchange.

I wish Sameer well and we part ways with an appreciated but slightly awkward hug and feigned kiss on each other's cheeks, as would be customary if we knew each other better. In many Arab circles, I had been treated as a true family member, so I knew what to do. The personal warmth that had been associated with business connections while working in the region has been a welcome surprise.

Moving on to some further interviews that will help me understand how the waterborne part of oil shipping has changed over the last decade, I attempt to speak to Ahmed Al Abadi, the CEO of the Egyptian national oil shipping company. He does not respond to a "cold call" email like Sameer did, so I convince Sameer to make a personal introduction for me, and the three of us agree to meet in a café to the side of the convention hall for coffee.

A man bumps me from behind. Ahmed is a tall, thin, and dark Egyptian man who speaks with a highbrow English accent. Ahmed manages distribution operations for downstream products, which meant the delivery of fuel from the nine refineries in Egypt to the storage tanks of the customers in other countries, from which gasoline and diesel fuel is trucked to retail filling stations. Abadi has Mutasim Al Nuaimi with him. Mutasim is the coordinator of inbound shipping activities at the shipping arm of the Egyptian national oil company in Cairo. I know Mutasim. I interviewed him a long time ago for a study my firm was doing for a private equity firm on accessorial systems for liquified natural gas (LNG) tankers.

According to Mr. Al Abadi, ocean oil tankering has changed in a few respects. Volume has been flat overall as renewables have been replacing fossil fuels and capturing a greater share of the total energy business (this is partly thanks to the fossil fuel tax). There are more LNG tankers for natural gas and fewer very large crude carriers (VLCCs) for oil. The cargo consists more of ethanol-based fuel blends, and less of pure fossil fuels, partly as a result of more labeling requirements at petrol stations that incentivized consumers to purchase clean fuels. There are more severe penalties for oil spills, gas leaks, ship engine idling, and violations of foreign flag rules that are designed to prevent tax evasion and abuse of labor by registering vessels in countries such as Liberia and Maldives. Environmental regulations promulgated by the International Maritime Organization have become much more stringent. Stainless steel or coated steel hulls are required in order to handle Group III lubricants. Ships must burn low sulfur fuel to meet Sulphur Emission Control Areas 0.1 percent requirements.[3] Carriers are required by law to treat ballast water with new onboard chemical treatment systems.[4]

Mr. Abadi says that there is less piracy and incidents are generally less damaging, as drones and Internet-based surveillance systems have permitted faster detection of potential intruders, and smart technology has enabled remote control of all ship operations in event of piracy incidents. As a result, risk insurance costs have decreased to their lowest level in more than a generation, crew hazard pay has been eliminated on most routes, and the armed security forces and forensic incident investigation business have returned to normal niches and have lost the celebrity sheen they had taken on during the peak of piracy.

Mutasim confirms Ahmed Al Abadi's analysis and expands on the importance of "clean ship" and "smart ship" technologies. An assortment of new technologies is reducing emissions, both "fugitive" emissions (from leaks) and emissions of burned exhaust, by controlling and optimizing ship engine idling and by detecting and minimizing gas leaks in the piping systems that run throughout the ships. Also, the ships now run on their own cargo fuel, the natural extension of a long-term trend which started more than two decades ago. Only a minority of LNG carriers used to be equipped with systems that allowed them to consume the cargo fuel for their own motive power. Today, all of them run this way, consuming 1 to 3 percent of their cargo in transit.

Mutasim briefs me on the way that smart ships have become a part of the internet of things. It has been quite revolutionary and the extent to which technology has influenced and optimized shipping to make it more efficient and green has been extraordinary, he says, but has not been noticed by the general public, which tends to be relatively uninformed about heavy industries such as oil shipping. Sensors and monitors collect data on every aspect of ship integrity and operations. The information is relayed to control systems via digital interfaces and protocols. Special software monitors ship performance and continually uses algorithms to optimize fuel efficiency and safety. The information system even collects data on structural integrity of the hull, which allows predictive maintenance and early resolution of incidences of corrosion and other damage. Autonomous and remote controls allow oversight from remote headquarters, which provides an additional layer of management supervision and security.

He also points out that receiving terminals had moved offshore to improve zoning and liberate land for commercial development in the city of Abu Dhabi. Offshore facilities are state-run and more economical to operate. In addition, the elimination of onshore liquefaction and regasification terminals has enlarged the amount of real estate available for commercial and residential waterfront development, which has made the city more livable. In fact, it has risen six places on *The Economist*'s annual ranking of livable cities.

I leave for the airport. The ride to the airport is quiet and relaxing. I take a combination of light transit link via a driverless subway, like at Orly airport in Paris, and a shared taxi ride in a plug-in electric vehicle, to the hotel. I take advantage of the forty-five minutes of a smooth ride and comfortable, quiet atmosphere to write my memo to Alex about the upstream oil and gas supply chain, one decade later. The airport is designed around human interaction and sharing of cultures, experiences, and emotions, in addition to human traffic flow. It features vibrant artwork like in the United tunnel in O'Hare airport in Chicago. The old airport nearer to the city center has been redeveloped into parks and recreational facilities. Bus connections to the city have improved, resulting in cleaner air, the faster transit to the city center, and a quieter ride.

At the airport, I pull out my ticket and receipt to find out which terminal and gate and need to be at. I notice that my ticket to Abu Dhabi is a bit more expensive than in previous years due to the carbon tax and to noise and pollution control measures, but I could have decided to buy a cheaper ticket: full disclosure rules allowed me to make an informed decision on the airline and the airport, including airport and aircraft air and noise pollution factors, CO_2 emissions, etc., and I bought my ticket from an airline with a lower D-Factor, which was not the cheapest one. On the plane everything is recyclable and is actually recycled, like food packs, headphone sets, cups, and glasses. The aircraft and airport operation is clean as a whistle because the government-sponsored Genuine Progress Indicator does not count spill clean-up and other revenue from socially nonproductive activities in the national income. In fact, the realization of the negative impact of certain economic activities triggered a range of tax disincentives for those industries. This was part of a larger transformation of the economic structure in Egypt, from a dirty industrial economy to a knowledge-based postindustrial one.

In the time that I spend checking in, winding my way through security, and having lunch once I reach the gate area, Alex looks at the improvement in upstream oil production due to the changes that occurred over the last ten years. Just as the flight is about to take off, I receive a reply email from Alex. It is my last Wi-Fi-enabled communication before I need to turn off my device.

He notes Egypt's conformance with a UN directive on carbon taxation, the country's transition from mostly natural gas-fired power plants to the current situation where nuclear plants supply over 40 percent of the energy for power generation, and a new mandate regarding ethanol fuel blends.

Overall Alex computes a 29 percent reduction in the D-Factor for Depletion, and observes that this had happened in a much shorter timeframe than would have been thought possible ten years earlier. He also notes that the national oil company preemptively diversified into renewable energy. The universities' example of 100 percent recycling and the country's coordination with the World Bank to achieve 60 percent recycling for durable goods had a demonstrable impact on the Depletion factor.

Alex is impressed by the human zoning (better mass transit, parks, and green spaces, etc.), the legislated reduction of vapors and fugitive emissions from tanker vessels, and the development of services industries and knowledge capital. Other petroleum shipping protections such as stainless steel or coated steel hulls, low sulfur fuel requirements, and the new law to treat ballast water with chemical treatment systems helped achieve the reduction. He computes a 34 percent drop in Devastation impact.

Respect for people and the human condition seems to have improved significantly, Alex said. He pointed to the country's public endorsement of the principles of the UN Global Compact, to the crackdown on foreign flag tanker registrations, and to broad scale adoption of D-Factor labeling. He also noted that upstream producers are taking steps to align their labor arrangements with global standards, in terms of technology transfer rather than slave trade. He also noted the improved and more transparent communications inside and outside the oil company, the influx of foreign nationals and young students into the country, and the decline in piracy achieved by surveillance, remote operation, and drone services. Overall, the Dehumanization factor dropped 24 percent.

On my way to Abu Dhabi I stop over in Dubai to enjoy a Desert Safari. My car, a Toyota Land Cruiser, is part of a caravan of seven Land Cruisers that trek several hours into the desert. We make one stop to dress up in Arabian Bedouin garb (I buy a fancy Omani sword from a touristy desert store for my son Brent). Then we merge with five other similar caravans until we were a group of forty to fifty Land Cruisers trekking further into the desert. We make one final stop to let the air out of the tires and let the engines cool off from the 140 degree F summer heat. Then the fun begins. We ride up and down the dunes, Arabian music blaring, holding for our dear lives onto the handles above our windows. The driver shows off his tricks, like driving sideways down the dunes, and driving headfirst down a steep dune, almost pitchpoling at the bottom. The wild and crazy experience is made all the more fun by the completely international group of people in the car. Most of them are from GCC countries, and a couple are from Africa. Everybody shares their culture with each other and learns a lot from the others. I leave for home on a high note.

My business is expanding in Abu Dhabi and I need to rent office space, so when I get to Abu Dhabi, I stop to see Shamsul Shahwani, who I had met at a conference in Bangkok the year before. Shamsul, or Shams as he prefers to be called, runs Al Wasil Shipping, so in addition to having office space it sounds like he may be able to offer some synergistic business connections for my company's consulting operations in the Middle East. Shamsul is a short man of Pakistani origin. He wears a full suit and tie, and his office looks well-worn, with dark wood, many small accoutrements and gadgets that he has no doubt acquired as gifts over the years. He has a secretary and an "office boy"—somebody to run errands, make copies, and handle administrative and bureaucratic tasks that seems to be such a necessary part of life in this region of the world. He is courting me as a customer for both agency and office leasing, so he takes me out to a very fancy fish restaurant on the Arabian Sea ("pick your fish and we will cook it for you"), and gives me a gift in a fancy giftwrapped box that I can tell is a desk clock from its weight and what seems to be a muted ticking noise inside of it. He proposes an agency agreement and shows me two office spaces that might suit my company's needs for the next couple of years. We part ways with a smile and a handshake, and our prospects for the future together look bright.

NOTES

1. King Abdullah University of Science and Technology. Wikipedia. https://en.wikipedia.org/wiki/King_Abdullah_University_of_Science_and_Technology. Accessed October 29, 2017.

2. "Human Rights, KSA and KAUST," by Adam. Jan 14, 2016. https://blueskybioinformatics.com/2016/01/14/human-rights-ksa-kaust/. Last accessed October 29, 2017.

3. "Ships Face Lower Sulphur Fuel Requirements in Emission Control Areas from 1 January 2015." International Maritime Organization. http://www.imo.org/en/mediacentre/pressbriefings/pages/44-eca-sulphur.aspx#.WgpwDkqWb7k. Accessed November 14, 2017.

4. "Fleet Growth Squeezes Crude Oil Tanker Market." BIMCO. https://www.bimco.org/news/market_analysis/2016/1222_fleet_growth_squeezes_crude_oil_tanker_market

SEVEN
Dammam to Marseille

PARIS, MAY 14, 2010

As part of an initiative to expand my company's presence within national oil and gas companies, I attended the National Oil Companies Congress in Paris.[1] The chants from the street were loud enough to disrupt my cocktail conversation on the eleventh floor rooftop of the hotel. The environmentalists were adamantly opposed to the National Oil Companies Congress taking place there. And during the speech the next morning that was slotted for the president of BP Tony Hayward, a heckler ran across the stage and grabbed the microphone, telling us all to stop drilling for oil. BP's Macondo well in the Gulf of Mexico was still spewing oil, even after nearly two months of attempts to "kill" the well. After the security guards clears the disruptor, the conference resumes.

After the conference, I took an extra day to explore how petroleum gets from the refinery to the filling station. Since my trip in France was going to involve extensive driving outside of Paris, including a visit to a refinery, I rented a car. Driving into the city for my first meeting, my throat became inflamed from high levels of air pollution. The city had put in place license plate restrictions to limit the number of cars in the city on any given day. Depending on whether your license plate ended in an even or odd number, you were prohibited from driving in the city on designated days of the week. The restrictions were not predictable, as they were in Bogotá. They were applied as needed, so you didn't know if the restriction applied or not. The driver needed to have a backup plan to take public transportation to be ready for unanticipated license plate restrictions based on actual air quality. It was annoying and a hassle for drivers, and it was an insufficient solution for pedestrians and residents. The smog was unbearable. I sought shelter inside while an attendant pumped up my gas, so I could benefit for a few minutes from ventilation that filtered out the pollutants.

I continued to follow the trail of gasoline at the pump by visiting my old friend François Valjean. François is a friend from twenty years ago. When I lived in France as an expatriate, François was one of a group of close friends who shared many dinners, parties, and weekends together. We kept in touch through our close circle of friends, and I knew he would be the right person to ask about the downstream oil supply chain in France.

François headed the ocean shipping and inland distribution division of a French oil company called PetroFranc. This division received refined oil (mostly gasoline) and made sure it got to the retail pumps. The company chartered tanker ships, took deliveries from third-party oil tanking companies, operated a network of receiving terminals and pipelines, and trucked fuel the last mile so all the filling stations have enough petrol to fill your car.

Talking to François was a little bit like talking to a CIA operative. He spoke with confidence and he seemed willing to share, but at the same time he knew so much confidential and proprietary information from his long executive career at PetroFranc that he needed to be careful not to go outside of the boundaries of public information for fear of disclosing stuff he shouldn't. Also, he tended to do business with a poker face that probably came from negotiating a continuous stream of high-value shipping contracts over so many years. However, he thrived on witty humor and intelligent debate about current events. He had spent many years abroad, especially in Dubai and Kuala Lumpur, setting up major midstream operations for PetroFranc's joint ventures in the Middle East and Malaysia. Since having returned to his native France, he has reintegrated into the corporate world headquarters.

"Where does the gasoline that was being sold at filling stations around Paris come from?" He began to open up. Depending on the bulk purchasing contracts, some of the crude oil originated in Russia, some from Saudi Arabia and Kuwait, some from Iran, some from Venezuela, some from Malaysia, etc. Most of the gasoline and diesel fuel came from refineries in France, Germany, the United Kingdom, and the Middle East. He knew far more about this than any consumer of fuel did. He was intimately familiar with the various origins of the hydrocarbons, such as the difference between Brent Crude and Arabian Light. I asked whether there was any way to know at the pump where the fuel came from. No way. There was no labeling or other way for a consumer to have any idea where the fuel came from. I asked if he thought there was a gap or problem. He said that no consumers cared about the origin of the fuel. They just wanted to fill up their tanks and get going.

"Are there clean fuel mandates?" The EU had just issued a directive the year before (2009) on clean fuels. The directive contained a provision that set a target of 10 percent renewables in transportation, and biofuels were one path toward achieving that target. However, there were no specific mandates on the production, sale, or consumption of ethanol-based fuels per se.

"Is there a carbon tax, and if so how does it work?" There was no carbon tax. In 2010 France was struggling with the concept of a carbon tax.[2] Energy production depended on energy consumption. About half of the cost of making oil is for energy itself, which includes oil, natural gas, and coal. Therefore, a carbon tax would be a "double whammy" to the oil industry, increasing the cost of inputs and reducing the sale of outputs. I asked him how he felt about the sustainability of this situation where an industry depends on Depletion for its existence, and any regulation would seem to limit or reduce its profits. Without any incentives to curtail the industry via carbon taxes for any other measure, it would seem that this situation would only end in complete Depletion of the natural resource. He answered as an economist. There is an optimal price, and if economically accessible reserves decrease, the price would rise, and the increased price would dampen consumption, resulting in a self-regulating system that will prevent complete Depletion. Also, he pointed out that technological advances were rapidly increasing the economically available reserves. The oil companies can be much more precise about drilling today than they could in the past, so their oil recovery rates are much higher, and the proportion of "dry holes" has decreased dramatically.

I couldn't argue with his logic about pricing, but that didn't address the issue of CO_2 emissions burning a hole through the ozone layer and causing global warming. "What about global warming?" I asked. "It's a hoax invented by environmentalist activists," he said. He referenced some studies that showed that global average temperatures over the last several decades were not rising.

Refocusing on questions of the supply chain, I asked how these large purchases of both crude and refined product were financed. He explained that the bank usually wrote a letter of credit to the oil company against the cash value of the important product. In other words, it would loan money to the buyer based on the collateral of the incoming product, and that would hand over the incoming product to the buyer when it received the money back. The letters of credit were sort of a very short bridge loan for these international transactions.

He had a previously scheduled lunch after our meeting with his banker, so he invited me to come. At a posh Parisian brasserie, I met Henri LeGrand de Genève. Henri lives in Switzerland and spends a few days a week in Paris for his business. Henri was a well-dressed, well-mannered and stylish man. Although he was in his sixties, he appeared much younger. His eyes moved in a shifty way that revealed he was always thinking something he shouldn't be thinking, but you never knew what that was. He struck me as a cross between Le Chiffre and Goldfinger, the James Bond villains. Henri was independently wealthy, thanks to a lifetime of bankrolling international oil and gas transactions. He lived in a $40 million home in the Alps. Once in a while he had to get on a private plane and take care of a customer in Mozambique or Egypt or Saudi Arabia, but most of his work could be done in Paris or from his home in Switzerland.

I came to know Henri LeGrand well over the ensuing years. He was a founding partner of an engineering firm a couple hours from Basel, and my firm partnered with his firm to execute a complex project in Latin America. I visited his company's office in Switzerland, and enjoyed lunch by a scenic marina while watching Henri's yacht set sail onto Lake Bodensee.

Nonetheless, I was beginning to feel like the downstream oil and gas supply chain had an unsustainable impact on the environment and society. The system of getting liquid gold from the ground into your fuel tank was well oiled. Many people made money on this supply chain, and many of them spent their whole careers in this business. It both reflected and contributed to the social, environmental, and economic fabric of society.

DISCOVERING DOWNSTREAM OIL

To get closer to the actual physical movement of the product, I took a trip to see the crude oil come into the country and get to the refinery. By leveraging some connections, I got an appointment to visit a refinery outside of Marseille. Marseille has a deep cultural heritage, and I appreciated it over a lunch of bouillabaisse with my friend Valentine Egbenedict, who I had met when he was managing the retail distribution network for TopUp, a major operator of the filling stations in Nigeria. Valentine had kept up with me via social media for years. On my birthday, there was always a Skype or LinkedIn birthday greeting from him. Randomly he would contact me via Facebook or email with a greeting, just to say hi. Quite a few people from Nigeria did this after my initial visits. I'm not sure if they did this for many people; I assumed they looked at me, with my education and credentials and American nationality, as one of their touchpoints to the developed world.

For less well-to-do people, Marseille is not one of the more desirable French cities to live in. It has a limited mass transit compared to Paris or London, and relative to other cities in France is corrupt, poor, and violent. There are drug problems. Furthermore, due to its proximity to North Africa, it has a long legacy of immigration and problems related to failure to assimilate.

Valentine had done well for himself. TopUp had relocated him to France, and within six months he had gotten a new job at a real French company, PetroFranc. He now managed PetroFranc's vast network of retail filling stations in France—a big step up for Valentine, and a feather in his cap. His job largely involved managing the real estate and retail stores that were required to support the filling stations. On a day-to-day basis, he planned store openings and closings, decided store assortments of products and merchandising formats, and handled the inbound logistics of fuel delivery to the stations, among many other activities that are required to keep the filling and convenience stations operating correctly and completely. He still wears the eyeglasses and many of the clothes from Nigeria, which are of a different style and quality level. Between that and having jet black skin, he stood out from many of the other PetroFranc managers.

I asked him about the process of clearing customs for bulk tankers coming into Marseille with crude oil. In contrast to the Saudi Arabian mystique of slow talking and slow-moving, Valentine is a fast talker and an active "doer" and taskmaster. His scope covers many geographies and he has hundreds of direct reports, so he must move quickly or else he will become irrelevant or obsolete. He always wants you to come to the point, quickly, before you finish your sentence. If your email is too long, he won't read it. He is an operations guy. He explained about all the paperwork he takes care of with each shipment: There were no electronic documents, which meant limited traceability in the event of inconsistencies or safety and security follow-ups related to irregularities like fraud, money-laundering, and terrorism. Some of the papers required included accounting statements, shippers outbound export declarations, verification of nationality and tonnage, verification of inspection, identification of passengers onboard, and a lot of tax documents to ensure that the government collects its share of the proceeds. There were also safety checklists to ensure that oil was not spilling from the vessel.

Once the product cleared customs, the crude oil was offloaded from the tanker into the oil terminal, which was adjacent to the refinery. To get a closer look at the refinery, I scheduled an appointment with Geoff de la Raffinerie, manager of the refinery. This facility, like most other refineries, power plants, and industrial factories, is disconnected from the social and commercial center of the city, and can only be accessed by car.

Geoff is an American expatriate who had been working for ExtonOil in France for the last five years. Originally from St. Louis, Missouri, he ran a series of petrochemical plants in Louisiana before running his first refinery. I met him while conducting an audit of his plant in Geismar, Louisiana. Geoff was my first real consulting customer. I had joined BLT Consulting as an analyst, and had been staffed on a number of projects as an analyst under the supervision of various project managers, but Geoff's company was the first one where my boss sent me off to do a project solo.

It was a great experience for me. Kudos to my boss, Bill, who entrusted me to do this. He knew I could excel at the technical parts of the project, but he also knew I was a green junior and could stumble in initial meetings where the client would expect me to have more experience than I did. So he shared some wisdom before I left for the airport. If I was ever unsure about what to ask in any meeting, just focus on five questions: (1) Describe your organization; (2) How do you measure your success?; (3) What are your biggest challenges?; (4) What projects are you working on?; and (5) What would you do to improve things? While I may have stumbled a little bit, Geoff was a nice guy who gave me slack, and was respectful of me as I spread my wings for the first time.

Geoff is 5′4″ and has a slightly protruding belly. By training, he is a chemical engineer from the University of Houston. He produced outstanding results at the other plants he ran, which is why ExtonOil chose to expatriate him to this plant. If he were not so good, they would have easily accomplished the same thing with a local French manager and avoided the hassle and cost of getting a work visa for Geoff and moving him and his family to France.

Geoff drew a stacked bar chart on the napkin. Crude oil is mostly carbon (about 84 percent) and hydrogen (about 14 percent). The rest consists of contaminants such as water, sulfur, metals, and sediment. Refineries process crude oil into a range of products (many for further processing) by pumping the crude oil through a boiler and heating until it separates by molecular weight at various temperatures. This fractionation separates the oil into liquified petroleum gas (LPG) (below 35 degrees C), naptha and gasoline (35–180 degrees C), kerosene and jet fuel (170–250 degrees C), diesel fuel (240–350 degrees C), and heavy oil (fuel for ships and thermal power plants) and asphalt (above 350 degrees C). A typical forty-two-gallon barrel of crude fractionates into the different grades of fuel (each characterized by different lengths of molecular chains) as follows: nineteen gallons of gasoline, ten gallons of diesel fuel, four gallons of jet fuel, two gallons of petroleum coke, two gallons of gas, and 5.2 gallons of liquified refinery gas, asphalt, lubricants, and kerosene.

Geoff drew on his seeming infinite knowledge to explain to me how refining produces many products I didn't know were petrochemicals. In addition to the main fuels, refining yields many chemical by-products, each constituting very small portions of the oil, which can be further processed to yield products such as ammonia, antiseptics, bubble gum, crayons, fertilizer, floor polish, heart valves, insect repellant, life preservers, liquid detergent, mascara, paint, sneakers, synthetic fibers, telephones, and plastic containers.

Making productive use of the heavy crude left at the bottom occupied smart minds for many years. It could either be coked (heated for a long period) to extract the remaining liquid, or cracked with catalysts to increase the yield of saleable products. Fluid catalytic cracking mixes this residuum and catalyst at high temperature in a cyclonic reactor, and hydrocracking used a two-stage slurry process to break it down. Reforming turned naptha into octane boosters for gasoline by restructuring the molecules and removing nonreactive components. Gas was separated from liquids (water and natural gas liquids [NGLs]), dried and separated, then shipped for sale.

Refineries were the third most energy-intensive industrial activity in the world (gas production and electricity generation are the first, and refineries use a lot of power). "In power generation, 64 percent of the energy used to produce heat and electricity was lost in the conversion process globally.[3] Combined cycle technology, the most efficient power generation technology that exists today, at best reached 62 percent efficiency, based on the newest, state-of-the-art gas turbines which were only starting to be sold. This means that 62 percent of the thermal energy stored in the fuel and the subsequent vapor that runs in the steam turbine could be converted into electricity.[4] After conversion, 6 to 10 percent of the power that is transmitted in overhead or underground power lines was lost due to 'line loss,' which increases with distance, and electricity theft."[5,6] Add to those inefficiencies an inefficient coal supply: in the United States alone, twenty thousand power plants railed in coal in dedicated unit trains of over one hundred railcars each, often from thousands of miles away, to fuel their boilers, day after day. By comparison, refineries for oil, gas, and petrochemical products, however, have much lower conversion losses than power plants.[7]

There was a colocated power generation plant dedicated to this refinery. From a distance, the power generation industry always appeared cleaner than the oil and gas industry, because the plants were usually separate from the mining and production of whatever fossil fuel is burned in them; however, the pollutants that resulted from the burning of fuels were more damaging to the environment than the oil, gas, and coal mining that provides the plants' fuel. Power plants collectively accounted for 24 percent of all pollutants from the burning of fuels. The energy losses in the production of electric power were astounding. Indirect emissions from the production of heat and power made up over 5Gt CO_2eq/year—roughly the same as chemicals, metals, cement, waste, and wastewater treatment combined.

Geoff interrupted our conversation to take a phone call. He talked about schedules and timings, and about some new pumping and conveyance equipment. He was talking to André Marchand, who runs the company that takes away the refined products and trucks them filling stations all around France. As the two were talking, I heard footsteps in the hallway. The door opened and a tall man walked in, talking on his cell phone. As soon as he walked in and Geoff and André saw each other, they both laughed and hung up their phones. "You didn't tell me that you were in the building," Geoff crackled. "I thought I would surprise you," cracked André with a mischievous smile.

André was a tall and powerful looking man with slightly deranged, crossed eyes that met yours and questioned everything about you before he opened his mouth, like a bear or an ape might do to size you up.

His fleet of trucks and drivers hauled the refined fuel from the refinery to the filling stations all around France. For nearby deliveries, he used his company's privately owned liquid bulk tank truck fleet. For farther deliveries, like to the Paris region, he subcontracted the loading of bulk railcars, and then the offloading to trucks for the "last mile" delivery when they got close to the delivery point.

Most of the trucks delivered to the stations in "flower stem and petal" route formations. Since the stations were evenly distributed, the most efficient way for trucks to deliver was by making one long straight trip (the "stem") from the refinery to the cluster stations, then stopping at each of the assigned stations until the truck is empty (the "petals"), and finally returning back to the refinery on the long "stem." Box trucks following deliveries and pickups followed a more complex and dynamic route, particularly by interleaving deliveries and pickups in the way that fills the most capacity in the truck, and by adjusting the route changing volume each day. However, with fuel deliveries, the fleet and the routes were much easier and much more stable and consistent because the truck had only deliveries, not pickups, and the stations have a very steady demand pattern.

André was well experienced in transportation operations since he had started his career in logistics in the army. Early in his career, he worked for a newspaper fulfillment company coordinating their nationwide daily deliveries. After that he held a variety of roles in the national telecommunications operation. He had a charming way with people, which gave him great success at process improvement efforts at large companies, of which he did several before getting invited to do the same at his current employer. The gig led to a full-time position, and he climbed the ladder over fourteen years to his current position.

Outside of work, André was a comic and an athlete. After the army he did stand-up comedy at a nightclub for three years, which honed his skills and made him able to crack anybody up. This talent was very helpful at defusing tension in important meetings, and helping people come to agreement on major issues. He invented words all the time—so much so that he was almost speaking a dialect. His inventions and mutations around language were clever and witty, and collectively really insightful. His special talent with language was a manifestation of the intelligence and ability to assemble, integrate, and connect all sorts of complex ideas. This made him a great manager and leader.

After having explored the oil supply chain from Africa and the Middle East to the French petrol stations, I was eager to get the downstream

D-Factor number from Alex. Alex came up with a total downstream D-Factor of 4.5 per tank of fuel. About half of that was related to Depletion. In bullet point formatting, he listed three things related to Depletion: (1) the energy intensity of oil refining; (2) the fact that the power plants that are needed to keep the refineries working were coal-fired and designed and built in a previous generation, which was inefficient to begin with; and (3) the extremely high rate of line loss electricity transmission and distribution for supplemental power that the plant generated which was not used by the refinery itself. An equal amount was due to Devastation, in which he considered the blight of the petrol stations scattered around the French countryside, and the fumes wafting from the filling stations into the nearby towns. A much smaller amount, 0.4, was related to Dehumanization, in which he considered Marseille to be a haven for crime and drugs, and the bureaucratic end to be an ineffective security screening process for vessels coming into the port.

I was curious if there was any way to set up this supply chain in a way that could minimize the energy intensity of the whole thing, and if the French consumers could have any influence over the working conditions at the other, upstream, end of the chain. A thought came to my mind: *if the French consumers knew* how much CO_2 was emitted in order to deliver the gasoline to the filling stations, and *if they knew* that more CO_2 was emitted by buying the Nigerian crude than by buying the UK crude (because of the Nigerian practice of flaring gas, which is outlawed in the UK), maybe they would be concerned about their transaction at the pump.

Another thought came to me later while I was working out at the gym in the hotel. *If the French consumers knew* that working conditions in upstream operations in the Middle East basically facilitated slave labor (or whatever you want to call it), whereas working conditions in upstream operations in Norway were among the best in the world, would they choose to purchase gasoline refined from Norwegian crude instead of gasoline refined from Saudi Arabian crude?

According to Alex, if French consumers would fill their tanks with petrol that comes from UK instead of Middle Eastern oil, this would shift more exploration and production activities under Western standards, increasing the energy efficiency of motors, pumps, and other apparatus used in drilling, and tighter standards for the disposal of contaminated water, including toxic chemicals used in fracking. It would reduce slavery conditions and assure that engineers and managers are covered by privacy laws, Western health codes, unemployment insurance, and pension plans. In all, it would cut the D-Factor in half. However, because of the cost of all these improvements, the cost per liter would increase by 35 percent compared to the current supply chain.

I dared to call François in Paris and ask him whether consumers would change their buying behavior. François said he understood the purpose and that the approach was admirable, but he pointed out the practical complexities of getting such an arrangement off the ground. He said that it would be hard to label the origin of the product because of the blending that goes on at the terminal before the refining operation, and some blending even occurs upstream before loading into the VLCC. Determining the origin with sufficient accuracy would require accounting at each step in the supply chain, and in the end, the label would probably have to list multiple sources with the percentage of each, as shirts are labeled "65% cotton/35% polyester." He also had doubts about whether consumers would respond to the information by changing their habits, especially if it meant paying more at the pump. Price competition at the pump is based on pennies of difference.

The natural extension of the labeling question was whether people would ultimately voluntarily pay more for one choice over another, potentially reducing their consumption, in order to mitigate the negative externalities in the global supply chain. "This is particularly unlikely to gain acceptance in the mainstream," he replied. He elaborated, carrying the proposition to its extreme (a classic French logical thinking process that is borrowed from Marxist theory): if they might pay 2 percent more, then why not 5 percent or 10 percent more? And how would they know the label is telling the truth, and that the company is not is trying to increase its profit margin by gaming the D-Factor label to charge a higher price? No, he said, people would only pay the lowest available price.

"What if the government mandated such labeling, as it has mandated nutrition labels for food, and content labeling for textiles and apparel?" "Hmm," he pondered for a minute. "Same answer," he replied. "Carry it to the extreme," he said, and you get the answer. From turning down the thermostat in the winter, to giving a handwritten letter as a birthday gift instead of a purchased retail gift, to a one-child policy, there is no limit to the extent to which government could intervene to reduce consumerism and its social and environmental externalities. However, government intervention to force such restrictive policies is politically dubious, as politicians who would suggest such policies would likely be voted out of office in the next election. I was well aware of this since Americans generally associate "energy conservation" with the perceived failure of such efforts during the presidency of Jimmy Carter, who was defeated after a single term.

This problem was indeed difficult to solve because oil and gas supply chains enabled everyday people to live their dream of mobility and independence, and made money for everybody along the chain, from the well producer and the subcontracted drilling company, to the banker and the oil tanker operator, to the marine terminal operator and the government that collected import duties on the incoming shipments of crude, the refiners, and the trucking companies, to the filling station operators. Any proposal that would favor, redirect, or constrain the free flow of oil would run into opposition from any or all of those parties.

Back in the hotel, thoughts swirled in my head as I packed my bags for my next stop, Ho Chi Minh City in Vietnam. I was exhausted and went to sleep quickly. My alarm clock went off at 5:00 AM and I managed to catch the first Réseau Express Régional (RER) train to Charles de Gaulle airport. Plenty of time to get through Security.

NOTES

1. For decades, national oil companies, which are state-owned, formed associations with international oil companies, which are traded on public stock exchanges, to acquire the technical expertise needed to explore and produce the oil underneath their soil. Since about the year 2000, that dynamic reversed as the national oil companies pursued a strategy of "resource nationalism," which consists of trying to retain as much of their oil wealth as possible by developing more technical skills domestically and aggressively restructuring their relationship with the international oil companies and oilfield service providers so as to require "local content" and force knowledge transfer to their own nationals. From 2007 to 2013, international oil companies' aggregate capital expenditures rose from $90 billion to $165 billion—a 183 percent increase, while international oil companies' collective capital expenditures rose from $150 billion to $600 billion—a 400 percent increase. The national oil companies with the largest investment programs were, in rank order: CNPC (China), Pemex (Mexico), Petronas (Malaysia), ADNOC (United Arab Emirates), PdVSA (Venezuela), Gazprom (Russia), KazMunayGaz (Kazakhstan), ENAP (Chile), Saudi Aramco (Saudi Arabia), and TOTAL (France). In 2012, PetroChina (China), Rosneft (Russia) and Petrobras (Brazil) raised $27b of debt, compared to $10b for ExxonMobil, Shell, PetroFranc, BP, and Chevron combined.

2. It will impose a miniscule carbon tax by 2020. See the Carbon Tax in France. http://www.planete-energies.com/en/medias/close/carbon-tax-france

3. Efficiency of Thermal Power Plants. Energy Efficiency Indicators. World Energy Council. https://www.wec-indicators.enerdata.eu/power-plants-thermals.html. Accessed July 14, 2017.

4. "7HA.01/.02 GAS TURBINE: The World's Largest and Most Efficient Heavy Duty Gas Turbine." GE Power & Water brochure, https://powergen.gepower.com/content/dam/gepower-pgdp/global/en_US/documents/product/gas%20turbines/Fact%20Sheet/7ha-hdgt-factsheet.pdf. Accessed July 14, 2017.

5. "Lost in Transmission: How Much Electricity Disappears between a Power Plant and Your Plug?" Wirfs-Brock, Jordan. November 6, 2015. http://insideenergy.org/2015/11/06/lost-in-transmission-how-much-electricity-disappears-between-a-power-plant-and-your-plug/. Accessed July 14, 2017.

6. David Jacoby (2017), *The High Cost of Low Prices*, Business Expert Press.

7. IEA (2016), Energy Technology Perspectives 2016 data visualization, www.iea.org/etp/explore/. Accessed July 14, 2017.

EIGHT

Abu Dhabi to Hanover

AMSTERDAM, MAY 9, 2030

On the way to Hanover, I connect in Amsterdam for my continuing flight to Hanover, Germany. After changing gates, I proceed to check in for the second leg of the trip. The KLM representative stops me behind a podium to ask me some security questions about my trip. "Where are you traveling from?" he says. "Abu Dhabi," I reply. "Did you pack your bags yourself?" "Yes," I answer confidently. "Were you in control of your luggage from the time you packed it until you arrived at the airport?" "No," I hesitate. "Actually, I left it at the hotel for most of the day while I was on a desert safari." Follow-up question: "Was it in a secure, locked area?" "No," I hesitate again. "Actually, it was in the hotel lobby. "Did the hotel keep it in a luggage room?" "No, it was not a locked or contained area. It was sitting right in front of the concierge desk the whole time—pretty much near the front door of the hotel." The airline representative nervously puts an "x" somewhere on the paper, but continues. "Do you have any liquids or other combustibles in your bag?" "No," I answer confidently. "Do you have any knives or guns of any kind?" I remembered the sword that I bought for my son Brent in the desert. Oh shit. "Yes," I said. "I have a large sword in my bag." This didn't sound very good. Now there are two black marks on my security profile.

The representative looks me in the eye, and begins to carefully assess the security risk that I might pose. I am a white American man on a business trip. I could see the gears in his mind turn as he decided to move on with the questions. "Did anybody give you anything which you packed in your bag? I remember the gift that Shams handed me his office. "Somebody did give me something." "What was that item that they gave you?" "Actually, I don't know what's in it. It's a box about 4 inches by 4 inches. I think it's a clock; at least that's what I assumed since it was ticking." "Who gave it to you?" The representative leans in and starts to become more keenly interested. "A Pakistani guy named Shams who I met in Abu Dhabi." I realize that this all probably sounds pretty sketchy to the security checker. "Did you know this person before your trip?" "No, I answer wimpishly."

At this point the security representative really has to figure out what to do with me. I'm sure the procedure calls for pulling me and my bag off the flight and detaining me for questioning in some far dark corner of the airport. This would be very disruptive for me and for the airline. The checker looks at his worksheet with all the black marks on it, then looks at me, then looks at the sheet again. Finally, he sighs and just says "Go ahead."

I am on my way to Hanover.

I go to the Hanover Messe (trade fair) each year to shop for suppliers of oil and gas equipment. The Hanover Messe, which began in 1947, is the largest trade fair in the world. It spans more than 40 pavilions, each of which is the size of many conventional trade fairs. The city takes in 225,000 registered attendees, more than 30 percent of which come from overseas, during a weeklong extravaganza each year. The city hosts a stream of other weeklong trade fairs dedicated to various specialty interests, but as far as I know the Messe is the biggest event of the year. More than 6,500 exhibitors present their products in categories ranging from Industrial Automation to Energy; Automated and Self-Driving Vehicles; Digital Factory, Motion Drive & Automation; Wind, Compression, Materials & Surface Technology; Industrial Supply; and Research & Technology. Many corporate buyers come to buy equipment. Schools make trips to the fair for the students to learn about cutting-edge technologies that are commercially available. It's not a futuristic visionary event, it's focused on existing technologies, but it tends to showcase the state-of-the-art of what is commercially available on the market. That is consistent with German engineering excellence—the focus is on world-class design and application, with an eye to the most useful, productive, and profitable applications of technology all the way across the spectrum. The broad scope of the event makes it interesting and important for anybody who is using or influencing the shape of technology and innovation as it applies to industrial products. It is fun as well as informative and useful. I wanted to bring my family for years, but the April time frame has always presented a conflict with my kids' school schedules.

Travel to Hanover always requires a connection somewhere in Europe and a short flight or a train from the connecting city to Hanover. This time I connect in Hamburg so I can visit a refinery there to see how the downstream oil and gas supply chain has changed since my visit to the refinery in Marseille in 2010. I know several people in Hamburg connected with importation, customs clearance, refining, and bulk transport.

This time I would drive from Hamburg to Hanover after those meetings. In good traffic it is only an hour and a half drive, but of course all bets are off during the week of the Messe.

Germany has implemented many changes over the last decade, many in conjunction with European directives on clean energy, clean power, recycling, and workers' rights. There is now an 18 percent tax on all fossil fuels, which has affected power generation, transportation, and industrial power. The EU issued Directive (EU) 2015/2193 limiting emissions of pollutants into the air from power plants between 1 and 50 megawatts. The directive regulates emissions of SO_2, NO_x and dust into the air, and it also requires power plants to monitor emissions of carbon monoxide. I call this the Clean Power Directive. The commission also set two targets for transport emissions: a 20 percent reduction from 2008 levels by this year (2030), and a 60 percent reduction from 1990 levels by 2050.[1]

Petrol is labeled at the pump according to its origin and type, its ethanol content, and its D-Factor. This enables consumers to choose more environmentally and socially sustainable fuel if they wish. Some studies have shown that about 30 percent of consumers consider the labels while making their choice of fuel.

The EU maintains its policies on working conditions, including working hours, part-time and fixed-term work, posting of workers, and informing and consulting workers about collective redundancies and transfers of companies, with the goals of achieving high employment and strong social protection, improving living and working conditions, and protecting social cohesion. These rules are minimum requirements; some individual member nations have tighter restrictions and mandates.

The EU has been consistent in its approach to "human" land use by zoning real estate on a mixed-use basis, as contrasted with the United States, which segregates land by its primary application. The US approach results in separate and distinct commercial, residential, and industrial zones. The premise of the US approach to industry is that commerce and industry are dirty and should be far away from residential land use to minimize the bleed-off of industrial and commercial ugliness onto human living. This premise, which evolved during the nineteenth century, is outdated, and loses an opportunity to encourage or even compel commercial and industrial operations to have a human footprint. This sacrifices the benefits of socially and economically cohesive microeconomies, and requires more commuting, which brings associated CO_2 emissions. Thus, European zoning incentivizes social and economic interaction at a local level, and the results are clear to see in many flourishing communities across Europe.

Consistent with mixed-use land zoning principles, the city of Hanover was upgraded and redesigned expressly for the Messe and for people who live there. It has become an international business center, well served by buses and rail lines up to 100 km outside of the city, in a way that generates income, growth, and technical expertise, all while bonding the residents together socially rather than segregating them by industry or activity. Access to the city is more or less radial, like Paris. Suburban and exurban communities develop equally around the central core. Human zoning may be even more evident in other European cities than in Hanover. Basel, Switzerland, is an international banking center, but it is also characterized by rolling hills, medium-density agglomerations, and inspiring architecture. Salzburg, Austria, is an intellectual, cultural, and musical capital that sits in a canyon in between mountains and is physically breathtaking. These and hundreds of other medium-sized cities have superb transport links and high-volume traffic flow, but without the ugliness that characterized their analogues twenty years ago.

REDISCOVERING DOWNSTREAM OIL

The mass transit system is easy to figure out in Hamburg, but just to make sure I stop in the information booth and consult an expert on my trip. The man behind the counter is about thirty-five years old and has wonderful interpersonal skills—not a low-level job, and not one to be disrespected, as would have been the case in many airport and railway station information desks in the past. The help that this person gave me made my day much better, and I am sure to show my respect and appreciation by taking the time for some greetings and small talk. He hands me a multi-modal printout showing step-by-step travel instructions broken down by every minute of my trip. Based on his advice, I take a combination of heavy and light rail from the airport to my first meeting instead of renting a car. Before going to the refinery itself, I stop to see a commodities trader.

I search the Contacts app on my phone for a fuels trader in Hamburg. I first started filing and updating my contacts systematically about twenty years before LinkedIn arrived on the scene. Over the years, my Contacts database has grown a lot. Now, with an iPhone in my hand, I search on "oil gas trade Hamburg." My phone searches thirty thousand records in an instant. The first name that pops up is Carmen LeFebvre. Carmen wrote me a letter applying for a job in 2010. Her credentials were very impressive, so I arranged to interview her in Paris. We met at a café off to the side of some public gardens in the fifth arrondissement. She was polite and polished, slow and deliberate—good qualities in a consultant. I offered her a job, but she decided to go into industry rather than consulting. She has maintained contact at career milestones, mostly through social media, and since we are in periodic touch, we are both happy to talk to one another again on the phone.

Carmen is now a fuels trader, about thirty years old. She is French but also speaks perfect German and perfect English. She is a consummate project leader who graduated from the Sorbonne and got an apprenticeship at BP in the United Kingdom, which developed into a good early career opportunity. Five years later, she took a more permanent and long-term career opportunity with DEA, a privately held German oil company, where she became a trader. Carmen was born of a French mother and a German father. Carmen is not meek or shy, but she always raises tone at the end of sentences as though they are questions, or she is seeking your confirmation. The trait is endearing.

Her trading activity has been affected significantly by the EU's and Germany's transition to clean fuel as a result of the fossil fuel tax, the Clean Power directive, and tightening vehicle emissions regulations. These regulations have shifted the whole economy, and DEA with it, not only toward cleaner fossil fuels but also toward solar and wind power technologies. DEA now invests heavily in solar and wind farms. Some of her colleagues from other departments have transferred to new projects in those areas because this is where the action is. For bilateral trades, and swaps, her customers have changed from oil companies and some conventional (thermal) power plants to a diverse mix of power plants and oil and gas companies, and a large share of the trades for all of those customers now involve natural gas and biofuels, as compared to oil in 2010.

This has shifted Carmen's supply base as well. She used to have a relatively small number of sources of oil—mainly Russia, Kazakhstan, Saudi Arabia, Kuwait, and Nigeria. Today some but not all of those countries supply the cleaner fuels that her customers want, so she has had to reach out to other sources. Her business has become more international.

Trading used to be based solely on price, but now that petrol is labeled by origin and type, ethanol content, and D-Factors, she has to buy many more different types and grades of fuel. An awareness campaign including advertising and product labeling has had major success. European consumers have been highly responsive to the option of buying cleaner fuels at a slightly higher price. They are willing to pay a price premium that is about four times more than American consumers to safeguard the environment—20 percent of the product price instead of 5 percent. Now filling stations stock more types and grades of fuel than before, and Carmen supplies them all. She even specializes to some degree in low D-Factor products.

The interview with Carmen really helps me understand how the oil and gas supply chain has changed. I thank her for her time and we exchange the traditional French goodbye kiss on both cheeks, which is unusual to do in Germany, but normal for us since we first met in France.

Once Carmen's work is done and the tanker company ships the product, the crude oil needs to clear customs. One of my company's board members is a freight forwarder and has an extensive network in Germany. I send him a text explaining what I'm doing, and he replies back with a cryptic message: Talk to Marco +49 40 326934.

I call Marco, who turns out to be a customs agent in the Hamburg port. Marco is twenty-eight years old and speaks good English, although with a heavy German accent. He graduated from vocational school with a specialization in information technology and programming two years ago, after doing his military service and doing an extended internship at a bulk shipping company. The internship was a worthwhile experience that helps him do his job better now that he needs to help shipping companies get through customs. Marco makes no attempt to be social. He doesn't say hello, and he doesn't usually sit with anybody at lunch, either. It's not personal. Once engaged, he is more open, but by his nature, he only responds to questions, so I have to ask him many direct questions to get the answers I am looking for.

It turns out that shipments and customs clearance have gotten a lot quicker, more accurate, and more reliable since I met with my friend Valentine Egbenedict in Marseille two decades ago. Ships register their cargo even before it is loaded onto the ship, and that triggers an advance notification to the customs authorities. The manifest is entirely electronic. The information flows into a pan-European database where analysts prepare accounting and tax bills, and screen for potential security and health concerns. Customs officials and independent standards groups test the fuel to verify octane and specifications before transferring custody to the power plants and retail customers.

Marco also tells me about how the EU uses Big Data to identify trends and patterns in shipments, which the governments use to adjust their strategic reserves and inventories of different types of fuels, and which the downstream oil and gas companies (including the retail filling stations) use to strategize their trades and inventory levels from bulk transport all the way up to stocks by grade at individual filling stations.

Once crude oil clears customs, it goes to the refinery, as before. For an update on refining, I contact my old friend Marcos Schmidt da Silva, who I met in Brazil while consulting to General Motors. Marcos is a very tall man (6′6″) with a slight humpback (which makes him look approachable, vulnerable in a small way) and a permanent smile. Marcos grew up in Germany but is half Brazilian. He speaks in a singsong sort of way, which I'm sure is due to his Brazilian side. This makes it easy to talk to him and makes him seem agreeable most of the time. He was educated in Cologne and joined BLT Consulting in the company's Brazilian office coming out of college. He rose through the ranks rapidly to become a partner, and secured a large contract with Wintershall, one of the larger German oil companies. After a successful two-year business transformation project, Wintershall invited Marcos to join full-time. Having initially joined as a regional sales manager for refined products, he rose to the position of vice president of the company's entire refining business unit. Marcos is a hub of social activity and is well connected. A combination of outlandish social parties and adventures, and a network of well-placed friends and colleagues, has helped him achieve considerable success in his career as well as growth for the refining business that he manages.

Marcos and I worked together years earlier in Brazil. We knew each other well, both at work—from working meetings to board presentations and team meetings in Sao Paulo, Chicago, and Düsseldorf—and at play—including fun in Brazil, including a team-building weekend at the beach in Fortaleza and sailing in Guarujá. It was great to see Marcos again.

According to Marcos, the refinery hasn't changed much in the last ten years. Refineries have complex designs that are based on enormous amounts of piping and vessels and valves designed to catalyze very specific chemical reactions. Once they are built they generally keep running as planned until they are decommissioned. This one is slated for shutdown in three years, since oil is declining as a fuel source, and the needed oil production can be shifted to newer, more efficient plants. The company analyzed the cost of retrofitting the plant, which would have extended the useful life, but couldn't justify the cost since the alternative was to just shift the production to another refinery with no capital investment. Hence the decision to shut the plant down.

There is one more piece to the puzzle to understand how the oil and gas supply chain has changed in the last two decades: the "last mile" delivery from the refinery to the filling station. For this part, I called my old client Ralf. Ralf is in charge of a consortium of tanker shipping companies that specialized in liquid bulk movements. Ralf is a tall, heavyset man who had black hair that was long where it grew, but it has left a bald spot in the middle of his pate. He previously started and ran a technical consulting firm in the freight transport industry. He sold that firm to a nationwide provider of logistics services to airlines and airports. After a three-year stint as the general manager to fulfill the obligations under the sale agreement, he took a position as managing director and significant shareholder of his current firm. With his technical consulting experience from his general management savvy, he has created many cost efficiencies and driven profit up. He has kept everybody happy. He is a family man with three children, and he keeps a close relationship with all of his relatives. He works a little bit too hard, and repeatedly promises to slow down to spend more time with his wife and children.

A client in Boston introduced me to Ralf more than a decade ago when Ralf was running his start-up company. Ralf and I met many times since, and our companies worked together on projects several times in the interim. Ralf was an ambitious and shrewd businessman, and when he committed to something he liked to do it right. I admired Ralf and appreciated this opportunity to interact with him once again.

"Hamburg's physical street infrastructure is designed for people, not vehicles, and trucks have been increasingly restricted from entering the city center," explained Ralf. The city is becoming more like Barcelona, with its active nighttime tourism and social life on Las Ramblas, or Amsterdam with its Leidseplein area and dense biking lanes, or Lisbon with its old narrow streets. This is good news for the residents of Hamburg, but presents challenges for deliveries to petrol stations. Ralf said they now need to make "graveyard shift" deliveries. And even then they need to work around noise restrictions, so this gives them only two delivery windows each day. To work within the new structure, they have worked with their oil company customers to enact several changes, including pipeline and drone deliveries instead of truck delivery, larger in-ground tanks with less frequent deliveries, and shuttles from larger stations on the periphery to smaller stations in the center of the city.

Over dinner I scribbled notes to send to Alex. Alex and I have by now developed a very efficient shorthand for communicating about the environmental and social conditions surrounding the operation of various supply chains. I send my notes and, on the other side of the Atlantic, he gets down to his calculations on the Downstream oil supply chain. Compared to two decades ago, the total Downstream D-Factor decreased by 44 percent, and the D-Factor for the entire (upstream and downstream) supply chain for tank of fuel went down by 31 percent.

Downstream, the reduction in emissions from both power plants, as a result of the clean power mandate, and from automobiles, as a result of the carbon tax, cut the Depletion factor for a tank of fuel by 33 percent compared to 2010. The phase-out of this refinery in favor of refineries that are newer and more efficient, in addition to the relocation of some filling stations to less urban locations and refueling via pipelines of truck traffic, and finally the increase in security that came with earlier and more accurate transmittal of information on inbound shipments to customs, decreased the Devastation factor by 46 percent. There was little change on the dehumanization side, except to note that several individuals in the supply chain changed jobs and careers, and are happier and more at the forefront of technology in their new roles.

Europe has taken a bolder position on environmental matters than the United States and developing countries. In contrast to the United States, where legislation can only prevent an activity after it is proven to be harmful, Europe takes a "Process" approach (the EU mandates the process and whatever product you make is fine so long as it respects the Process). The Process approach assumes that the action is harmful until it is proven safe.[2] This leads to a much more cautious environmental approach.

Even back in 1992, Denmark enacted a carbon tax that applied to all consumption of fossil fuels (natural gas, oil, and coal). In 2001, Germany made a landmark decision to phase out nuclear energy by 2020. In 2001, the UK instituted a climate change tax on natural gas, electricity, petroleum and liquid gas, coal, lignite and coke. In 2005, the EU established an Emission Trading System (ETS) that included all large CO_2 emitters such as power plants, large industrial installations and platforms for oil and gas production. In 2007, the UK passed a resolution mandating 20 percent renewable energy by 2020. In 2008, the UK passed another Climate Change Act that set a conditional target to reduce emissions by 50 percent by 2027 (on the condition that the EU adopt similar reduction targets post-2020). The same year, the UK targeted a 34 percent decrease in greenhouse gas emissions by 34 percent by 2020. In 2012, the EU included CO_2 emissions from aviation in the ETS, Denmark set a goal of 35 percent of energy from renewable sources and 50 percent of electricity from wind. That same year, Bulgaria, France, and the Netherlands 2012 banned fracking. In 2013, Denmark set another target to cut greenhouse gas emissions by 40 percent by 2020 compared to the 1990 level, and decided to phase out oil for heating, and coal by 2030.[3] In 2016, the EU ratified the Paris accord which was consistent with all the previous pronouncements, directives, and targets.[4]

I feel revitalized after seeing all these dear friends again after so much time. My next set of appointments is in Kuala Lumpur (KL). The trip will take fifteen hours and involve stops in Munich and Singapore. In addition to the travel time, Kuala Lumpur will be seven hours ahead of Hanover. I think about how I will need to power through my presentation in KL without falling asleep!

NOTES

1. The Medium Combustion Plant (MCP) Directive. European Commission. http://ec.europa.eu/environment/industry/stationary/mcp.htm. Accessed October 29, 2017.
2. D. Vallero et al. (2012), "Products," *Unraveling Environmental Disasters*, Elsevier, Oxford.
3. David Jacoby, "Comparative Economics of Combined Cycle, Solar, Wind, Hydro, and Geothermal Power." Presentation to the University of Calgary, Haskayne School of Business, Calgary, Alberta, Canada, October 19, 2017.
4. For more detail on European environmental legislation, see the Appendix: A Brief Chronology of European Environmental Legislation.

NINE

Marseille to Ho Chi Minh

HO CHI MINH CITY, JUNE 2, 2010

Viet means people, Nam means South. The Chinese controlled Vietnam for one thousand years. This is why the language uses Chinese phonetics. The South belonged to Cambodia and the North was Chinese. The French controlled the country from 1858 until 1954. Ho Chi Minh established the Communist Party in 1933 and declared Vietnam's independence from China in 1942. The Vietnamese fought the French and the French lost at Dien Bien Phu in 1954.

The Americans invaded in 1955, and made the North capitalist north of the 17th parallel. The Americans controlled the entire country for twenty-one years, from 1955 to 1975. When the Americans left, they organized a civil war, for which the Vietnamese still hate the Americans. Americans lament their body count from the Vietnam War, but about fifty thousand Americans died in the war, compared to two million Vietnamese. After the Americans left, the country became Communist from 1975 to 1986. Then it became Socialist in 1986 . . .

> Tunnels. Tiny access portals. Too small for Americans to get in. Wood door sealed to keep wood smell from Americans. Cramped. Doubling over. Big juicy centipedes. Pitch black. Drops and rises. Hot. Stifling. No air. Miles. Couldn't do it; would go insane. Soldiers went insane. Traps. Snakepit trap. Armpit traps. Crushing torso traps. Americans fighting Viet Cong, and Vietnamese fighting Viet Cong. So Americans killing the wrong enemy and Vietnamese killing other Vietnamese, sometimes the wrong ones. Vietnamese used rubber tire sandals to identify the Viet Cong; people wearing them were imprisoned or executed.
>
> —Excerpt from notes of my June 2, 2010, tour of the Cu Chi Tunnels

In 2010 Vietnam was a paradoxical mix of capitalism and communism, where a string of normal-looking things was punctuated by the weirdest, most abnormal things. The stock market was taking off, but the houses were all the same dimensions because the government dictated the plot size and the house dimensions. There was allegedly free expression, but the government speaker, who was the keynote speaker at the conference I went to, spoke for only five to ten minutes and then refused to take any questions, leaving a big void in the evening's agenda. There was a flourishing tourist trade, but the housecleaner in my hotel room stole money from my backpack (I caught him with his hand in my pack!). Traffic was congested, but all the drivers needed to slow down and veer to the left to avoid a dead man in the right lane. My tour guide says the body was fake, but it looked very real to me. The list goes on. The longer I stayed, the more weird things happened.

I had two commitments in Ho Chi Minh City. The first was to deliver a speech on "Global Economic Trends and Containerised Trade Growth Development," to a group of transport and port development specialists. The Vietnamese economy was growing at 7 to 8 percent in 2008 and 2009, and there was a sense that the country was positioned to play a much greater role in international affairs, owing in part to its recent nomination to be a nonpermanent member of the UN Security Council.

Shipping costs were low, but service was poor. In order to compete effectively in the shipping sector, the ports needed to offer more services and improve their service:cost ratio. These companies also needed to be more flexible and innovative. With public sector support, the country had the potential to envision the completion of several long talked-about regional freight corridors. One was the East-West Corridor connecting the South China Sea, across Vietnam, Laos, and Myanmar to the Gulf of Martaban. Another was the Asian Highway—a network of roads stretching from the Indochina Peninsula including Vietnam up into China and westward through India and Bangladesh toward denser networks in Eastern Europe and the Commonwealth of Independent States. A third was the Trans-Asian Railway, which had a similar footprint. Since many of these projects were multigenerational and depended on commitment from other countries, I was suggesting that Vietnamese port developers focus on water links to Singapore and Manila, and road links to Bangkok. This would connect them to three regional trading hubs.[1]

I also planned to attend the posh three-day Wharton Alumni gathering at the Sheraton Saigon Hotel, which was going on the same week. A keynote speech by Ambassador Charlene Barshefsky titled "The U.S. Response to Globalization and the Rise of Asia" caught my eye. Also, regional alumni were slated to talk about their companies and projects. Investment bankers would be sharing their interesting perspectives on Vietnam and investments in global markets. Some government ministers and prominent faculty were also on the agenda.

DISCOVERING BOTTLED WATER

Over cocktails at the Wharton conference, I met Chi Pham. Chi, an investment banker, wore a life of hard work and hard-won successes on his face. His face showed a maturity that most Western faces will never know. He would never take his success for granted, as he knew what it was like to be poor, and he was undyingly loyal to his family and business clan for their support during his rise to success. He always wore a dark suit out of respect for the people who have helped him throughout his life and career, and respected everybody, even and perhaps especially those who had hurt him. As a child, Chi lived through the French occupation, Chi never saw his parents or either of his brothers or his sister again after the communists took control in 1975.

Chi and I sat at lunch, eating our chicken and drinking our bottled water. I noticed that the bottled water we were drinking was a French brand. I wondered if that water had come into the country on the same flight that I was on from Marseille to Ho Chi Minh city the previous day. No, that was highly unlikely. It was probably transported by ship from Le Havre to Hanoi, and then redistributed within Vietnam to Ho Chi Minh City. If that was true, I thought of the true price that was being paid for me to sit in Vietnam and drink bottled water from France. If a single passenger's transatlantic flight caused the same amount of environmental and social damages as drinking 1,124 cups of coffee, then shipping bottles of water from France to Vietnam must have been a horribly damaging thing to do. Bottles of water were much heavier than coffee, and the bottling didn't allow for efficient packing since the bottles are round and leave a lot of empty space between them. I was advised to drink bottled water during my stay in Vietnam, but I wondered if there was a more environmentally sound way of doing it.

I chatted with Alex about my curiosity, and he told me that the bottled water supply chain was a topic that was getting a lot of attention in those days. Everybody seemed to be coming to a big "'Aha!' moment" about the craziness of shipping bottled water around the world, and were beginning to connect the dots to globalization and question everything about global supply chains.

I met Chi for coffee the next day, and asked him about bottled water from France, since Perrier, Evian, and other brands of bottled water from France were very popular around the world. Chi didn't know much about this but, as I had suspected, he knew somebody that did. Chi introduced me to the general manager of a bottled water company, Nguyên Tràn. On a brief conference call that was half in Vietnamese, Nguyên invited me to the aqueduct the next morning. Nguyên was a businessman with interests in a wide range of activities. One of them was running an operation that produces bottled water for export.

The next morning Nguyên picked me up at my hotel and we drove to his workplace together. Nguyên Tràn, a short, thin Vietnamese man. Like most Asians, he was not overweight from eating processed food with additives like most Americans did. By American standards (where obesity was becoming the norm), he looked downright malnourished. He spoke very quickly and without heavy accent, making everything he said sound simultaneously fluent and incomprehensible.

Although the domestic bottled water market was dominated by foreign companies such as Nestlé and its locally branded products, Vietnamese companies produced and exported bottled water, too. He took me on a tour of his operation and explained how it worked. First, he said quietly, as if I was not to repeat it to anybody, the company was a subcontracted operation of Eau Yes!, a leading French beverage company. It imported bottles under the Eau Yes! brand name, and exported bottled water from a variety of Vietnamese sources under the brand name Tropic Sunshine. He did not explicitly say why the relationship was hush-hush, but as the day progressed I realized that it was probably because Eau Yes! would not want its brand name connected with the working conditions, the safety, or the environmental footprint of this operation.

After a visit to the Mekong Delta, with its "muddy" waters, and a couple of hours at the War Museum, where I learned in graphic detail how Agent Orange and other chemical weapons devastated Vietnam and created birth defects for three generations, I had concerns about drinking water that was originally from Vietnam, regardless of whether or not it was in a bottle. After my speech, I would have the chance to take a tour of the Mekong Delta, during which a restaurant served very fancy sailfish. However, after seeing the habitat of this fish, I had no appetite for it.

Along the way we picked up one of his employees (I couldn't repeat his name if I tried, and we didn't communicate much because he spoke no English). Workers live in nice area but drove old cheap diesel cars to work because there were no public transport alternatives.

The springs were in a mixed-use area that included steel works, a large number of small Asian-style retail shops, and apartments. The patchwork of property interests evidently had no zoning: some units had power and some didn't. Some had water and some didn't. There were no green areas, and there was no urban center. No permitting was required to operate the aqueduct; water was a free natural resource. The power grid was unreliable, so noisy and dirty diesel generators provided backup power.

As we toured the pumping area, I noticed that old, poorly maintained pumps were leaking oil into the ground. The staff moved around the property in kerosene-fueled cars. Kerosene lamps added to the pollution at night. Loud cars and machinery screamed. People tossed garbage onto the ground and weekly garbage was piled in heaps instead of buried underground to avoid the cost of disposal. The experience reminded me of an experience I had on the road between Delhi and Rishikesh, India. There was an elaborate religious ceremony on the banks of a river. The people conducting the ritual were dressed in beautiful golden satin robes and burgundy headdresses. The sun was setting and the scene made a beautiful photo opportunity. But shortly before the sun set some organizers lit kerosene lamps by the shore, and the golden-robed people in the ceremony were also carrying kerosene lamps and waving them around as part of their dance. The ceremony and the backdrop was beautiful, but the air in the general area was smothered in dark plumes of toxic kerosene smoke that hung low in the air, neither rising nor falling.

The government had passed some laws about safety and environmental standards, but no fines had ever been collected: a 10 percent "tip" kept the police away, a $5 bribe staved off a $50 penalty, and so on. Successive administrations allowed the practices because they personally earned more take-home pay this way than if the money went into an official bank account. This meant more cash for their cars and TVs and mobile phones. The government had no industrial plan, no published goals, and no key performance indicators. In a democracy the leadership would have been voted out, but in a communist regime, the lack of evidence of governance failures actually made it even more unlikely for change to occur.

The level of education and training was generally low. According to the Hays Global Skills Index, no Asian countries came in high on the ratings for skills development, skills versus job matching, or sustainability of higher wages.[2]

As I may have expected in a corrupt political environment with a general lack of education among the workforce, working conditions were subpar. No doubt some of this arguably came from the workers themselves, who are not educated enough to establish safe working environments, and some was due to the practices of their employers. Nearby Bangladesh (later) became infamous for poor workplace conditions when the Rana Plaza building in Dhaka collapsed, killing more than 1,100 garment workers. Many workers were intimidated and mistreated, especially if they tried to organize a union. Factory owners hired thugs to beat the organizers. Some owners allegedly threated to kill the dissenting workers. Although a law protected workers, violations were common. Managers and owners frequently intimidated the workers so they would back down before reaching the 30 percent threshold to register a union.[3]

My tour of the springs covered the upstream part of the supply chain for bottled water. Now I needed to go farther down the supply chain to get the whole picture. The next morning at breakfast I met Friedrich Werner, a German businessman who I had encountered at the port conference. Werner knew everybody in Vietnam. Literally, I think he knew everybody. And he was into agriculture and food trade. He even started a coffee plantation business.

Werner introduced me to Rodney, a middleman who bought up water and merchandise of all kinds and loaded it onto containers for export to other countries in Asia and to the United States. Actually, Rodney worked for another middleman, who work for another middleman, who worked for a large, global freight forwarding company.

Truckers picked the water up in small pickup trucks and carted it to regional depots, where it was aggregated so larger trucks could haul it to the port. It went to the old port, which was close to the downtown area. The convergence of heavy industrial logistics involved in loading the ships, and the heavy foot traffic related to downtown commerce and retail trade, made for a chaotic and awkward mixed retail and industrial downtown experience.

Rodney walked me through the cargo area of the airport where imported bottled water entered the country. On our way to the receiving area, we walked through an area that was like a retail mall for domestic appliances. According to a taxi driver who took me back to the hotel later, Robbie and other middlemen were not just trading water for export to Asia and the United States. They were profiting at the fringe of a human trafficking corridor between Asia and the Middle East. The appliances were sold mostly to ladies returning from work in the Middle East. Saudi Arabia had domestic (household) labor agreements with Bangladesh, India, Pakistan, Philippines, Vietnam, Sri Lanka, Tanzania, and Niger. Indonesia, Nepal, Vietnam, and Cambodia were also working closely with the Kingdom to sign agreements.[4] According to the taxi driver, the recruiters looked for poor girls from remote areas. They were hired under legitimate contracts for domestic services, but some of them got sucked into abusive domestic situations, sex slavery, or sex trafficking. Anyway, according to the labor agreement between the countries, the ladies received subsidized housing when they returned from three-year labor contracts in the Middle East, and when they resettled they needed household appliances.

Rodney said his responsibility for imports of bottled water ended after the water was received by Customs. After that a local distributor trucked the bottles to stores. To follow the trail of the water, I called my friend Tony Liu, who I had met at a similar ports conference the year before in Bangkok. Tony managed a Chinese freight forwarding company with a branch in Ho Chi Minh City and in many other cities throughout Southeast Asia. He coordinated multimodal transportation shipments for all sorts of companies—consumer products companies, textiles and apparel companies, electronics companies, automotive parts suppliers, you name it. Tony was a young, thin Chinese man who wore tailored suits, well-polished shoes, and narrow ties. Conservative was a good style for the transportation and logistics business in Asia. Most customers didn't expect you to innovate, they just wanted you to ship the product and get there on time. Tony used to be a computer programmer before getting into transport and logistics. This served him well since most of the advances in shipping and freight forwarding the last couple decades were related to better, more complete, more timely information flow, which allowed the goods to be tracked and traced as they moved around the world in various forms of containers and on different vessels. Tony had an entrepreneurial business on the side in transport technology—an optical reader solution that will make moving containers reports more efficient and less costly.

Tony explained the transportation part of the bottled water supply chain, and since he saw both exports and imports, and he relied on both to balance his loads, he described both aspects to me.

For imported bottled water, the bottles of water were unpacked (in the maritime business, the technical term is "unstuffed") from the container and loaded onto small Asian urban delivery trucks that hold fifty to one hundred cases each. Some of them were covered and had sides, but many were just flatbed trucks that carry cargo that is stacked two to four cases high. The trucks zigzagged through crowded city streets, dodging scooters, bicycles, and pedestrians. There were 45 million motorbikes in Vietnam, and 7.4 million in Ho Chi Minh City itself. Only 8.4 million people lived in Ho Chi Minh City—including babies, children, and the elderly—so most people had more than one motorbike.

Driving any vehicle—motorbike, car, three-wheeler, or whatever—through Ho Chi Minh City was a wild experience due to the crowded streets and the many different kinds of vehicles all around you at any given time. To begin with, the route was not clear. All sorts of small shops selling cheap plastic stuff, toys, engine parts, electronics, a little bit of everything, spilled out into the street. There was no organized parking, so motorbikes were everywhere, many on the sidewalk. This made no room to walk, so people walked around the motorbikes. Every driver and passenger wore a face mask due to the pollution—both direct pollution (from their own motorbike) and pollution from the general level of smog due to the collective impact of all the motorbikes). Safety was a joke: people were carrying all combinations of crazy stuff on their tiny motorbikes, like oversized gasoline jugs and children, and both of those together. Almost nobody wore a helmet.

Some bottled water was also exported. For this trade, urban delivery truck drivers gathered small boxes of full bottles and loaded them onto container ships. The container ships sailed in a circle from northern Asia down to southern Asia, over to Mexico and up the American Pacific Coast, into Canada, and back over to Japan and Korea. Each container ship was a mammoth creation, hauling stacks of containers piled twenty high, twenty wide, and eighty from front to back—32,000 twenty-foot-long containers per vessel. The containers were stuffed with everything from jellybeans to auto parts. Only the largest ports with the deepest drafts could accommodate these leviathans, and ports had supersized their capacity to compete for their business. Container shipping began in the 1950s when an American named Malcolm MacLean started shipping goods in metal boxes from the United States to Puerto Rico. Until that time, every kind of merchandise was loaded onto general cargo ships in bags or crates or some other nonuniform unit of containment. Loading and unloading the ships was labor-intensive and took forever. As metal boxes became the standard, those metal boxes were adapted to move directly from the container ship onto trailer chassis, container shipping became a whole new business and global trade took off like wildfire.

In order to understand where the exported bottled water was ultimately consumed, I talked to Arthur Waterman, head of Business Support for Asian operations at I-Shop (before I-Shop divested of its Asian operations). I had interviewed Arthur four years earlier for a white paper I wrote for the Economist Intelligence Unit on the future of retailing. Arthur handled imports for I-Shop, which was a large, international retail chain focused on food and grocery stores. The company made a major push into Asia during his tenure, and during this time he was responsible for importing bottled water into I-Shop's stores around the world.

Arthur was tall and stocky, and spoke English with a distinct and light accent that I could not place. It must have been Dutch because he reported to a boss in Amsterdam and his biography on the company website, which I checked before our meeting, said he had studied marketing at Eindhoven University. His brown hair, brown eyes, and pale skin stood out in Asia, where he was based while on an expatriate assignment for five years. Arthur had a talent with interpersonal relations, especially with foreigners. He was good at managing teams, and in Vietnam he had found a place where his skill was greatly valued, since importing from and to everywhere in the world takes an extraordinary team effort. He was good at troubleshooting problems, improving processes, and making teams perform better.

Arthur lit up a cigarette, as he always did before, during, and after meetings, or at any time you put a drink or food in front of him, as if he needed a cigarette to prepare for the impending feast. He explained the consumer decision-making process at the point of purchase in the United States. Consumers in New York bought their water mostly based on the shape of the bottle and the design on the label. Despite the fact that the bottles were shipped nine thousand miles over sea and land, it was not clear to consumers where any of the water came from. They paid $2.50 for a normal bottle because that is what they would have paid for an alternative drink like milk or soda, and up to an extra $2 per bottle for a nice design and form factors such as a pretty picture on the label and a curvy bottle that was easy to hold. The upper bound was $7 for a high-energy protein drink, which provided a lot of headroom for pricing of the other drinks. Plus, water was perceived as a healthy drink, which was statistically worth a premium because two-thirds of the customers were obese. There were no recycling cans at the restaurant, so all the bottles went directly into the trash, where they were collected weekly by garbage trucks and sent via rail, truck, and barge to a landfill in Bolivia.

I had gotten all the information I needed to share with Alex. Alex crunched the D-Factor analysis for bottled water based on the information I gave him and some research of his own.

Compared to drinking water from a tap or from a nearby source, it was hard to imagine a more intensive abuse of hydrocarbons. The empty bottles were manufactured in China from polymer chemicals and shipped to France (for imported water) and Vietnam (for exported water), then full bottles of water were trucked and shipped on containerships halfway around the world. Ocean and rail transport over six thousand miles, carrying one of the heaviest cargoes, generated intense exhaust gas composed of nitrogen, water vapor, and carbon dioxide (CO_2). For Vietnamese exported water, there was no permitting required and anybody could take and sell water. The pumping station leaked oil. Multiplying the cost of each stage of this value chain by that country's ranking on the Global Green Economy index yielded a Depletion factor of 0.52 per bottle, said Alex.

The plastic bottle manufacturing process emitted toxic chemicals into the air. Consumers dumped their used bottles into trashcans that went into landfill the next week when trash trucks picked them up with all the other refuse at the curbs in front of their houses. The infrastructure, from the pipeline to the cargo ship to the delivery trucks and the trash trucks, blighted the landscape from Marseille to Ho Chi Minh City (for Eau Yes!), and from Ho Chi Minh City to New York (for Tropic Sunshine). Multiplying the cost of each step of the process times the inverse of the Environmental Progress Index yielded a Devastation factor of 0.63.

The social Dehumanization factor was the about half the value of the Depletion and the Devastation factors. The employees in Vietnam worked without protections or standards. For some bottles that were manufactured offshore (Alex assumed the plastic bottles were manufactured at a petrochemical plant in China), those workers operated under discriminatory laws. Some female workers who became pregnant were fired. Some overstressed male workers committed suicide, according to a human rights watchdog group. The Dehumanization factor accumulated to 0.25 across the whole value chain. Overall, the D-Factor was 1.4. That was fairly close to the D-Factor for a cup of coffee.

I asked why Arthur didn't buy the water that was sold in Europe and the United States from a closer source. Why didn't consumers in Vietnam buy spring water from Vietnam itself, or from Cambodia or Thailand instead of France, if it needed to be purer than Vietnam's native sources? Why didn't consumers in New York buy water from Upstate New York or Vermont or Maine instead of bottles of water from Asia? This would have reduced the transportation cost (and the CO_2 emissions proportionately) from 31 percent to 20 percent of the sale price, and opened up the possibility of reusing the bottles, which would have eliminated significant amount of nonbiodegradable landfill. The price would have fallen by 14 percent because of the reduction in ocean shipping of heavy liquids, and the manufacturers could have pocketed some of the difference in higher profit margins.

The D-Factor would shrink by 2 percent, said Alex, as France's worse scores on greenhouse gas (GHG) emissions and environmental performance mostly offset its better scores on the Social Progress Index. France emitted an average of 27 total GHG emissions including land-use change and forestry ($MtCO_2e$) in 2012, while Vietnam emitted only 12.[5] I didn't trust that figure because I knew from experience that the air in Vietnam was problematic, but then I remembered that I had trouble breathing the last time I was in Paris, too.

I asked Arthur if I-Shop or anybody else had considered eliminating this dirty supply chain altogether by reducing consumption, either by changes in behavior or values. He replied, of course, that it ran contrary to I-Shops' financial interest to curtail consumption. "The shareholders would not be very happy with that decision," he said. However, he did concede, off the record, that more and more water was being bottled locally or being bottled in bulk rather than in small bottles. I said I would monitor that trend.

As I prepared to leave Vietnam, I couldn't help but think that some Asian governments had chosen to develop economically first and implement environmental and social policies "later." Environmentally, China was far behind that of Western Europe or the United States. Almost 76 percent of China's energy came from coal-burning power plants, compared to the 35 percent global average.[6] Left alone, the trend would have gotten a lot worse due to their young and growing populations and increasing consumption patterns as more people joined the middle class.

I needed to get to Shenzhen to work for a transportation company there, but I had a few days in between. As there was not enough time to return to the United States, I opted to spend the weekend in Sri Lanka with my friend François, the bulk oil trader for PetroFranc. François flew in from France and met me in Colombo.

Our ecotourism excursion was breathtakingly beautiful and selfish in its unsustainability. We saw mongooses, crocodiles, water buffaloes, painted storks, meerkats, spotted deer, peacocks, leopards, elephants, green beater birds (which looked like hummingbirds), wild boars, and gray monkeys. However, our tour of "Block 1" of Yalla National Park became intolerable after the third hour. The kerosene exhaust and the clattering noise from the engine and the vehicle were hard to take. We couldn't hear the birds over the noise of the jeep. The jeeps were made by the Sri Lankan subsidiary of a major Indian brand. If this vehicle was anything like the trucks I helped a large German truck maker to manufacture and export from Brazil to Malaysia, they were probably designed to be easy to assemble locally, and cheap to buy and operate. Assembling them in Sri Lanka undoubtedly created local jobs and reduced import taxes, but resulted in lower overall quality and low or no emissions standards compared to jeeps that were exported as a complete vehicle from an advanced economy. It seemed that these tours were doing more to destroy the nature preserves than to preserve them.

The lack of respect for the animals was manifest not only in the vehicle but also in the lack of education of the tour guides. They compelled the peacocks to move off the road in front of us, disturbing the natural balance, called to the deer, and threw rocks at the animals to make them scatter so we could see them move. I can't blame the tour guide—he was being entrepreneurial, making money for his family, and helping the tourists enjoy their visits. But with two hundred of these jeeps crawling over the place all the time, I felt guilty that we were spoiling the animals' habitat with diesel fumes, noise, and human intervention. I couldn't help but think how much of a difference electric vehicles, even golf carts, would have made to that whole experience.

I asked the tour guide why he used kerosene-powered vehicles. We had a lot of trouble communicating because he didn't understand the question. Part of the disconnect was his weak knowledge of English, but another part of it was that he only knew kerosene vehicles. He had never owned a diesel or gasoline-powered car or truck. Eventually I was able to understand that kerosene fuel was sometimes (but not always) a few pennies less than gasoline or diesel, which made those jeeps more attractive to them. In addition to the tangible outward scars like exhaust and noise, a low level of awareness and education made any near-term solution—indeed any voluntary, even in the long run—solution to this problem difficult to envisage.

After the weekend, I changed gears both physically and mentally to get ready for my trip from Colombo to Shenzhen the next day. The scheduled flight was a direct one, and only five hours long. I looked forward to some tasty Chinese food as I set my alarm clock, pulled the quilt up to my neck, and let myself doze off.

NOTES

1. David Jacoby, "Global Economic Trends and Containerised Trade Growth Development." Presentation to the 6th ASEAN Ports & Shipping Conference. June 5, 2008.
2. NDTV (2015), "Government to Train over 40 Crore People under National Policy for Skill Development by 2022," http://www.ndtv.com/india-news/government-to-train-over-40-crore-people-under-national-policy-for-skill-development-by-2022-780165. Accessed June 15, 2017.

3. HRW (2014), "Bangladesh: Protect Garment Workers' Rights," https://www.hrw.org/news/2014/02/06/bangladesh-protect-garment-workers-rights. Accessed June 15, 2017.

4. "126 Saudi Recruitment Offices Stopped from Recruiting from Eight Countries." Al Arabiya English. http://english.alarabiya.net/en/business/economy/2016/10/31/126-recruitment-offices-stopped-from-recruiting-from-eight-countries.html 2/. Last accessed October 6, 2017.

5. David Jacoby (2017), *The High Cost of Low Prices* Business Expert Press, 2017.

6. IEA (2014), "World Balance 2014," www.iea.org/sankey/. Accessed June 14, 2017.

TEN

Hanover to Kuala Lumpur

KUALA LUMPUR, MAY 20, 2030

Based on my previous experience in Southeast Asia, I assumed that Malaysia would be a land of sweatshops with slave labor producing plastic toys and household goods like rubber dish-drains and wastebaskets for American stores like Walmart. I thought it would be underdeveloped—dirty, poor, and cheap—like Cairo or Shenzhen during my 2010 trip.

Boy, was I wrong. The natural environment is lush, green, and tropical. There are palm trees, birds, and blue lakes. The country is rich. Not glamorous like Dubai, where every building has to be a monument to outdo the one next to it. But clearly comfortable, happy, and well-off. Plus, there are plenty of astounding buildings in Kuala Lumpur to ogle at, including the Petronas (Twin) Towers and the KL Tower (the one with the ball in the middle, like the Oriental Pearl Tower in Shanghai).

The Malay language is funny and cute. There are many words like bing bing and pata pata pongi. I chuckle when I see them written on signage, and laugh when I hear them spoken. Actually, the city name is pronounced differently than I thought—it's not koo-A-la lum-POOR, but rather Kwala Lumper (spoken quicker like two syllables instead of five, and emphasis on the Lump). But everybody says KL for short, anyway. They speak English much better than in China or Thailand. It's a totally English-speaking place, so it has attracted a lot of international visitors and trade.

There is a blend of food. Not just Malaysian, but also Indonesian, Indian, and European. I walk past dozens of Thai and Indonesian restaurants.

Compared to the United States, Malaysia seems to be a "land of normal proportions," like the United States was when I was a child, before the steroids made their way into processed foods and people got fat and decided to drive monster trucks instead of sedans. The people are about "the right" size and the cars are European-like.

I am in KL for the 36th annual ASEAN Ports and Shipping Exhibition and Conference. Malaysia is on one of the busiest shipping routes in the World on the Strait of Malacca, and KL is a major port city. The conference is held at the luxurious five-star Renaissance Johor Bahru Hotel. The first day consists of a tour of the Ports of Pasir Gudang and Tanjung Pelepas. The next days are a nonstop frenzy of high- and low-level ceremonial and informal meetings with government officials as well as more technical marine professionals such as harbor masters, port engineers, shipping companies, and freight forwarders.

As a VIP and sponsor, I am invited to an early morning assembly in a royal room with the senior officials from the government's transport ministry as well as the heads of the largest companies in the conference. All the local officials are wearing their ceremonial local dress, which is flowing with silk and gold everywhere, very colorful. The VIPs sit in gilded chairs that are arrayed along the edges of the room, looking at the others across the room. There is nothing in the middle. This preconference schmooze is primarily a photo opportunity and a chance to shake hands with the most elite people in the conference. I exchange words with the VIPs sitting next to me—some chitchat and explaining something about my company. They clearly don't understand what I am saying, nor am I following what they say, but we are all smiling anyway. Everybody is more focused on looking good than on doing any business here. And this is just the preconference meeting. The official start of the conference is a twenty-five-minute ceremony of pomp and glory for the local transport ministry officials.

On the first night I participate in a VIP dinner. There are about twenty-five of us seated around a long table, and I am seated strategically next to one of my target customers, thanks to the forethought of the conference organizer, Pei Chyi Ling, a thirty-five-year-old Malaysian marketing expert at TransExpo. On the second night there is a spectacular Gala Dinner which is choreographed to give us a chance to mingle with a more broad group of conference delegates.

On the third day of the conference I duck away from the technical sessions to present a proposal to Nagasa, a large industrial conglomerate in Malaysia. Most of the companies I consult to are relegated to ugly, remote industrial areas, so I am pleasantly surprised that Nagasa is located in a campus of green trees and grass, and there are birds chirping as I walk down the path to the entrance of the main administrative building. A huge grassy area with a set of soccer fields sits between the administrative building and the next building over. My proposal is related to a training program, and it encourages me that an industrial manufacturer based in this beautiful setting is focused on knowledge development and learning.

Over breakfast I read in the newspaper that Malaysia adopted the UN's system of Environmental and Social Performance Indicators. In conjunction with the performance indicators, the government has developed a social, economic, and environmental five-year plan. Citizens have rallied around the plan, both for and against. Overall, citizen engagement has increased considerably. Town hall meetings occur with some frequency, and are well attended.

The unfamiliar juxtaposition of a meeting with an industrial company in a lush green environment reminds me of a meeting I once had at Maschinenfabrik, a German company whose factory in Friedrichshafen, Germany, was situated along Lake Constance (the "Bodensee" in German). To get there, I flew into Zurich and drove along the edge of Lake Constance, then crossed the lake in a car ferry. The dramatic scenery of the lake, the mountains, and the surrounding small towns and roads was heaven on earth. One of my prior trips to nearby St. Gallen, where I was presented with a prize for an essay contest and spoke on a panel with government leaders and industry titans about European integration, left me with a similar sense of awe and admiration for how the Europeans can make manufacturing so human and clean.

Returning to the conference on the third morning, I run into Pei Chyi again. She seems to know everybody, and is of local origin. On this trip, I am refreshing and updating the information I had collected in 2010 on my trip from Marseille to Ho Chi Minh, to update my study of the bottled water supply chain, I ask Pei Chyi if she could introduce me to somebody in the bottled water industry in Malaysia. She connected me to Phang Hoàng, who runs a bottled water company in KL.

REDISCOVERING BOTTLED WATER

Phang picks me up the next morning at my hotel. She is courteous and professional. She supervises a production operation for one of the largest exporters of bottled water in Malaysia, called Song. The company had only local operations when she was young, but has recently begun export operations. A short, outspoken woman, Phang had a thick Malaysian accent when she spoke English. Her accent is so strong I ask her to repeat every sentence at first, but after trying to understand everything she says, I settle on understanding what I can (about 40 percent) and not worrying about the rest.

Her great-grandfather founded the company in 1926. She grew up in a small village and moved to KL for college. Phang loves social media and posts everything she does or says to Facebook and Twitter and Instagram and Snapchat. She makes at least eight posts of us together during my brief time with her.

We drive to the aqueduct together. If I were not visiting today, she would have taken the tram to work. Most of the employees of Song live in a mixed-use residential area with a park in the middle, which serves as a meeting place for the community. Musicians play in the square in the evenings in their children's activities and weekends. A light rail transit link carries them to within a short walking distance of work.

A new agreement between the government and the bottled water company requires the company not only to acquire the land on which it travels from the water, but also to pay per unit tax on the use of water as a natural resource. This is an extension of the carbon tax law that applies more broadly to all natural resources.

Phang shows me the process flow, the workers, and the equipment. The cleanly dressed workers represent a mix of men and women. About ten years ago Malaysia passed legislation to improve maternity leave, to make all employers offer crèche services, and to limit overtime. Workers drive electric golf carts around the facility. The pumps are all the latest generation, most energy-efficient ones with the motor inside a can so oil cannot leak out. These pumps are now cheaper than less efficient ones thanks to subsidies provided by the government to encourage energy efficiency under a program coordinated by the World Bank. They were originally subsidized by the government, but have been commercially successful without subsidies since they became popular enough to produce a high volume and benefit from economies of scale in production.

The local company has an exclusive agreement with StarWater, which is now the largest bottled water company in the world. After traditional bottled water companies took a lot of heat in the press for their unethical and unsustainable global management practices, StarWater moved in and bought most of them and began operating under a different business model. It bottles locally, uses recycled and reusable containers, and requires its partner operations around the world to follow the management practices and principles prescribed by the UN Global Compact. In addition, it discloses all these relationships in its annual report in a new section mandated by the Carbon Disclosure Project. For StarWater, which has operations around the globe, this "compliance" section of its annual report is three pages long. Since changing its business model, StarWater's stock price has tripled, and tens of thousands of tourists have flocked to Song Springs.

Overall, this processing operation and its ecosystem make for a much "tighter ship" and a cleaner and nicer place to live and work than the bottled water operation I toured in Ho Chi Minh ten years ago. Like black and white.

To update my knowledge of the distribution channels for bottled water, I ask Phang to introduce me to one of her distributors. As we enter the loading dock area where the trucks are loading crates of bottled water on the trucks, she places a call on her mobile, talks for about two minutes in Malay, and then puts her phone back in her purse. She tells me we will have to wait about fifteen minutes, so we sit down for a cup of coffee. Soon, a man shows up and introduced himself to me as J.J.

J.J. informs me that the bottled water business has changed a lot over the last ten to twenty years. Almost half the water sold in bottles does not come from springs anymore, but instead from local taps. It is bottled and distributed locally, but the international supply chain is gone, except for labels and other minor supplies. The increase in local refilling has transformed the distribution and shipping from a multimodal, exhaust-spewing ordeal into a simple, relatively clean urban delivery network. Distributors make a lot more money because there is a large degree of brand confusion between water that really came from a spring far away, and water that was just filled and distributed a few kilometers away. They profit from this confusion to make relatively high profit margins on simple bottles and jugs of plain municipal tap water.

Another change is that the growth in the business has slowed to the point where it is now almost level from year to year because consumers are conscious of the environmental impact of global shipping. The government imposed a "greaen labeling" rule that made retailers declare the source and purity levels of the water. Every bottle now contains a mandatory D-Factor label, which states the origin and purity of the water, and quantifies the social and environmental externalities of that product supply chain. It is analogous to the Windows Experience Index that Microsoft (ah, remember the old days!) used to represent in a single number that reduced the complex interactions between the chips speed, the memory, the hard drive technology, and the graphics card, among other factors into a single number. These labels have encouraged small businesses to provide purer water with less negative effects on the environment, and at lower prices than the exotic imported water.

J.J. also points out that the practice of bribery went away a few years ago when it was discovered that some large bribes were funding drug traffic. The confluence of inner-city gang violence and drug traffic led to the election of a strongman who cracked down on both. Several corrupt high-ranking politicians were thrown out of office, and extensive media exposure of the event has discouraged corruption ever since.

I am curious whether and how the transport link of the supply chain changed, so I call to ask Tony Liu, who I met in Ho Chi Minh City nearly two decades ago, and have run into several times since. Tony travels around the region and knows every country well. Tony doesn't work in Vietnam anymore, so he refers me to Yoong Nim Chee, who manages a local cartage firm in Kuala Lumpur. Yoong and I meet for coffee that afternoon. Sipping a cup of coffee, I see him pull up in a new Tesla.

We chat for a few minutes about his new car—the new Model Z—then he explains that all the delivery vehicles are also battery-powered. There are recharging stations every few blocks throughout the city, so within the city limits there is no shortage of motive power. Most of the electric grid has been replaced by household and small commercial site-based energy production through solar panels and fuel cells.

The new delivery vehicles are quieter and generate no exhaust. Music from street musicians flows through the air instead of the sound of combustion engines, and many of the people who had spent most of their lives wearing air filters over their faces finally walk around with their faces exposed. Diesel, kerosene, and gasoline are banned as fuels for passenger vehicles. All the residents that drove these cars traded them in for more fuel-efficient petrol-driven cars, or for all electric cars, under a trade-in program subsidized by the government. The government's trade-in program stemmed from a United Nations climate change incentive to get more people driving clean fuel vehicles. The UN worked with the World Bank to create a global fund to accelerate the transition from petroleum-fueled cars to alternative fuel vehicles.

The air is now clean as a result of the elimination of petroleum-fueled cars and practices such as filling them with kerosene to save some pennies. The spotted owl, which was on the verge of extinction because of air and water pollution, is off the endangered species list and is starting to propagate again. Now that most of the country has been geocoded and mapped, and internet access is widespread, all the delivery trucks use Google and other routing and mapping products as a matter of course. This makes delivery more efficient, especially considering that the integrated real-time traffic information is embedded in the routing solutions.

Half of the new delivery vehicles are self-guided and self-driving. They are basically automated bots on wheels, delivering to customers who pay by holding their debit cards within range of radio frequency identification readers, and the bot places their water on the sidewalk of the preconfigured address.

All of Malaysia uses real-time traffic information, combined with wireless and contactless payment systems, to keep traffic density to a minimum. The whole area was a pioneer in the technology. Its directly adjacent neighbor, Singapore, was the first country in the world to introduce an electronic congestion pricing system back in 1998. Singapore's Land Transport Authority, which has responsibility for all modes of transportation in the country, oversees implementation of intelligent transportation systems in Singapore. Singapore is a world leader in electronic road pricing. It has used congestion pricing in its city center since 1975, although initially the system was based on a rudimentary system of license sheets in windscreens and spot-check police enforcement. In 1998, Singapore implemented a fully automated electronic road pricing system that used Dedicated-Short Range Communications. In-vehicle units installed in each car still accept a prepaid stored-value smart card called the "Cashcard." Singapore's ERP scheme uses traffic speeds as a proxy for congestion. It adjusts the price of tolls to optimize traffic flows along a speed-flow curve. The optimal speeds are set to 45 to 65 kmph for expressways, and 20 to 30 kmph for arterial roads. In effect, the system uses market signals to manage supply and demand.

I ping Arthur Waterman, from the retail chain I-Shop, on Facebook, and find that he is in town to monitor compliance to the company's new workplace standards rules and regulations, which were put into place along with Malaysia's new environmental rules. We both follow the moving dots on each of our screens, each representing the other, until we literally bump into each other outside a Starbucks.

Over a café crème, Arthur informs me that new rules in Malaysia dictate that every retail store and restaurant must have recycling bins and they must give patrons discounts for recycling their bottle or can and arrange for proper disposal of the material in the bins.

Nonrecycled trash has dwindled to a trickle of what it used to be, thanks to the creative energy and activism stimulated by organizations like the Ellen MacArthur Foundation, which work as an agent of the EU and the UN in incentivizing waste-free and recyclable product design. Today products sold at retail outlets are nearly 100 percent recyclable.

I was quite impressed by all the changes since my previous trip in 2010. I asked Alex to do his magic. He started off by explaining how hard it was to model the change because of how totally different things are today from 2010, but I finally got the answer from him: Overall, the D-Factor for bottled water shrank by 67 percent. A massive, global initiative sponsored by the UN away from plastic bottles and toward local bulk packaging of water has completely eliminated the type of plastic bottles that used to populate landfills and float around the ocean. This has had a double benefit—on the one hand, the reduction of energy-intensive and toxic petrochemical manufacturing operations, and on the other hand, the near complete elimination of plastic bottle waste. Even the energy which had been expended in recycling the plastic bottles was saved, since the bottles were no longer produced in the first place. With some adaptation of his old spreadsheet, Alex is able to update the numbers to today's reality. The D-Factor for Depletion fell by 73 percent compared to the Vietnamese bottled water supply chain of 2010.

The bottling operation has been adapted to a production facility for under-sink water purifiers for residential and commercial use. Breakthrough efficiencies in sand filtration, active carbon filtration, reverse osmosis (ultrafiltration), mineralization, and sterilization have had secondary benefits in industrial processes as well. The Devastation impact has shrunk by 66 percent.

Low-skilled jobs in truck transportation, warehousing, and convenience store retail sales have been eliminated by the reduction in volume of physical flow through retail supply chains. A larger number of jobs in information technology, artificial intelligence, robotics, and energy storage has offset the loss in basic jobs. Child labor has been nearly completely eliminated, since the skills needed for the new economy were much more advanced. The Devastation factor has fallen by 59 percent.

After the Ports and Shipping conference was over and my tour and interviews for the bottled water supply chain were complete, I think about how, ever since I was a child, I have yearned to visit islands in the South Pacific, especially Fiji, Tahiti, and Bali. Unfortunately, a prebooked itinerary impels me to start my trip to Beijing, where I have a schedule packed with appointments. I depart, hoping that someday I will have the chance to win a huge multiyear consulting project on a tiny island in the South Pacific!

Boarding the flight to Beijing, a diverse and multinational assortment of stewards and stewardesses of various ages greets me as I enter the plane. Ten years beforehand, some airlines used to run recruiting operations to specifically import young Malaysian ladies between eighteen and twenty-four years old to be their flight attendants, according to a Thai airport operations director at a conference in Bangkok. The girls had to fit within weight limits, have smooth beautiful dark skin, walk and talk gracefully, and smile a lot. Today, the crew still looks sharp and young, but new functional requirements and a new objective selection and grooming process channels the best recruits into the best positions. As a result, the whole crew has excellent capabilities and interpersonal skills, and they learn leadership skills that they are portable to many other jobs. Every one of them is fascinating to talk to.

ELEVEN

Ho Chi Minh to Shenzhen

SHA TIN, NEW TERRITORIES, CHINA, JUNE 9, 2010

When I checked into my hotel in Sha Tin, an industrial town in the province of Shenzhen, the hotel reception handed me a message in an envelope. I was invited that night to a special dinner as the guest of honor. The ceremony began with preliminary greetings. People were introduced to me and said things to me, but since they only spoke Chinese, I didn't know who they were and I didn't know what they said to me. I smiled a lot. Dinner was lobster in the shell. It was an exotic dinner, but the only tool for eating it was chopsticks, and for as hard as I tried to figure out how to get through the shell with the chopsticks, I fell short. I used a nutcracker and a pick, and luckily the host supplied a bib and wipes. I saw that other people were getting some lobster meat into their mouth, but it looked like a very steep learning curve. There was karaoke, or at least some kind of singing show on a stage. The main singer looked and sounded about twelve years old. Was she related to somebody in the company? Was this legal? Was it going to be my turn at some point?

The next morning I attended an executive board meeting. This meeting was on a high floor of an office building with a panoramic view of the New Territories from a huge glass window—I felt very powerful standing and staring out the window, like a businessman on the cover of *Forbes* magazine. Several people introduced themselves. The meeting started. There were twelve Chinese client executives around a large boardroom table, and they were all speaking English. However, as the meeting went on, they rapidly accelerated their pace of speaking, and the comprehension that I had had with these people prior to the meeting dissipated. Within ten minutes they were all speaking what seemed to me to be "Chinglish," and I didn't understand a word. The meeting went on for an hour. I tried to take notes, but ended up with less than a sheet of paper. It looks like my best bet at this client was going to be one-on-one meetings.

I was working for the Canton Freight Transport Corporation. Canton is the old, British, name for Guangzhou. My client was a short, stout man with short black hair and a small bald spot, named Dick Zhào. Dick ran operations, which mostly consisted of funneling incoming shipments from the industrial areas around Guangdong into Shenzhen and through to Hong Kong for reexport. It was responsible for the whole multimodal operation, which often started with rail and truck shipments, included warehousing and storage, as well as loading and unloading railcars that shuttled freight between Shenzhen and Hong Kong. A lot of the operation was manual and labor-intensive. It could have been mechanized, but given the cheap cost of labor and the relative lack of skills required for the job, it was more economical to use people than machines.

Dick always smiled a lot and he was very successful. Maybe he smiled a lot because he was very successful, or maybe he was very successful because he smiled a lot. I couldn't figure out which. He seemed to only work between about 10 AM and 4 PM each day. (Maybe that also explains why he smiled a lot.)

DISCOVERING SMARTPHONES

Smartphones were a raging success in 2010, and the vastly popular one was made in China, so I decided to take advantage of the fact that I was in China to learn how they were made and shipped to the United States. Dick, my client, knew how everything got everywhere. He explained the dazzling movement of raw materials and components around the world. Smartphones were made of hundreds of components. Silica was mined in China for the microchips, which were made in Japan. Lithium oxide for the batteries came from China. Bauxite was mined in Australia for the aluminum housing. Petrochemicals for plastic components came from South Korea. Wireless modules came from India, and electromechanical systems came from the United States. The assembly plant in China integrated this vast, globally sourced set of materials. The units were shipped from Hong Kong on Australian freighters to all destinations for sale in retail department stores.

To tour an actual production plant, I needed a further contact. Two years earlier my father had connected me to an American Chinese man working for an electronics manufacturing company that made the smartphones in China. The man, Tony Tan, generously offered his time and insights for a white paper I was writing for *The Economist*, and *The Economist* published my interview with him. Tony had graduated from Yale (Math) and Brown (Engineering), and he was enjoying a prosperous career supporting Chinese subsidiaries of US companies, especially in the area of machined mechanical and electronic products, since these manufacturing companies were enjoying an unprecedented rise during the 2000–2010 period. Playing the role of a pundit on the burgeoning role of Asia in US companies' supply chains, I had written dozens of articles and delivered dozens of presentations on the topic, many under the title "The Asian Sourcing Boom: How Long Will It Last?"

Tony was delighted to invite me to visit his plant outside of Shanghai to talk about the smartphone supply chain, so I started my journey from Guangzhou to Shanghai. After arriving at the Shanghai airport, my trip to the factory consisted of two taxis and a bus. The trip reminded me of an adventure I had years earlier with a junior consultant who wanted to experience the way "real people" traveled between Shenzhen and Guangzhou (Canton). From our hotel in Hong Kong, we took a taxi to a bus to a train, then we hitchhiked to another bus station where we took a second bus to a taxi to the hotel in Guangzhou.

As I traveled to Tony's plant, I noticed that there was an unpleasant odor—not fresh or clean. I vaguely remembered that smell from East Germany in the 1990s. Maybe it wasn't a bad smell, but rather the absence of good smells—no grass or flowers, no laundry detergent or cleanser, no fresh paint, no pastry or coffee. A stroll through Paris would bring all of these scents, but not here. I also noted utilitarian architecture that was a bit depressing. There was barbed wire in many places to keep people out—even in places that would seem to be public. The bus seatbacks had big Chinese writing on them, like advertisements or state-sponsored messages. Slogans on highway overpasses and in the airport hall read: "Achieve excellence, but pursue the unsurpassable" and "Strive to be good; long to be great." It seemed a bit strange.

Shanghai was like six cities in one, with each city containing a full range of different zones and areas and neighborhoods. Each quadrant of the city, if you will, consisted of an enormous industrial city, a huge commercial and trading city, and a retail city, and housing apartments were built on top of all of them. Due to the incredibly fast growth of the country and of Shanghai, I considered opening an office there. An extremely nice realtor showed me some apartments and some small offices. There were plenty of adequate facilities, but the rents were outrageous—higher than New York, Paris, or London.

After a lot of fumbled communications with the taxi driver, who spoke only Chinese, I arrived at the factory, which was hard to find because it had no sign outside. Advertising wasn't necessary, or even desired, because the company was operating under a subcontract, and its name never appeared on the packaging or in any publicly available documents in the West.

The plant consisted of a manufacturing clean-room, a wing for other component subassembly operations, and a main electronics assembly operation. At first glance, the production and assembly operations looked like factories in Europe or the United States. Based on batch whole systems, workers moved small piles of in-process inventory to the next workstation, helped along by visual displays showing the number of units needed and the number of units left in the in-process inventory pile. However, while the workers learned basic quality concepts such as "5S" (sort, set, shine, standardize, and sustain) and root cause analysis ("fishbone diagrams") that implied they would be able to produce the same products as Western plants, they fell far short of implementing "Six Sigma" quality. Six Sigma is a process control standard in which 99.99966 percent of all production events are statistically expected to be free of defects. This equates to 3.4 defects per million. Manufacturers that fail to come close experience much higher product failures, and consequentially have higher rates of returns, as well as inventory imbalances. Basically, they can make a product that looks the same but does not function as well (or at all!).

Other components were made inland, so I took a trip to see one of the inland factories in Hexian, a city in Anhui Province. I flew from Shanghai to Nanjing, then took a two-hour car ride to the factory. Door to door, the 363-kilometer trip took about seven and a half hours—basically the whole workday. I arrived in the evening. The contract manufacturing company had arranged the hotel, so I expected things to go smoothly. However, there were problems with the reservation when I arrived, and nobody spoke a word of English. The room was basic—one step up from camping. There was no restaurant, so I ordered room service, but when I looked at the food on my hotel room desk, I lost my appetite. I went out for a walk that evening. Apparently the town did not do much urban planning—there was a huge factory and some wide boulevards and small retail outlets of various kinds here and there. A discontinuous sidewalk came and went as I walked in and out of the street. One of the retail-looking stores, which had an open door and some lights on inside, looked like a cross between a gambling shop and a peep show. The people and the human spaces for interaction were miniscule compared to the imposing footprint of the factory and the boulevards.

My tour of the Hexian factory was tightly coordinated and brief. They were very busy producing as many components as they could as fast as they could. Based on the interactions that I observed inside the plant, stress levels were very high. As if I were his buddy, Tony shared some inside perspectives with me. The employer maintained all the leverage in the relationship, and this resulted in some pretty perverse outcomes. Ladies who became known to be pregnant would be fired so they would not become financial drags on the company. Men who had limitations or mental illnesses would be pressured beyond their limits, and some would harm themselves or even commit suicide. There were no support mechanisms, and in the event of mistreatment, there was no acceptable mechanism for employee recourse against the employer.

The factory got its silica and lithium oxide from mines in China, and because China relied heavily on imported scrap to manufacture its wares, it imported a lot of contaminated scrap. Other countries discovered that China was a fertile dumping ground for contaminated waste scrap, so they shipped all of theirs to China. China Customs knew that if it stopped the flow of imports to China, most of the exporting countries would simply divert that cargo to other Asian countries that have even lower standards,[1] so it allowed the imports. While Organization for Economic Cooperation and Development (OECD) countries cut industrial discharges of heavy metals and persistent chemicals by 70 percent to 90 percent,[2] more than 70 percent of industrial wastes in developing countries in Asia were not treated before being discharged into water.[3]

To better appreciate the inland logistics flow, I returned to Shanghai on the truck that was consolidating electronic components and spare parts, and hauling them to the assembly plant. Since the Hexian plant was the first on the route, we started in a small low-rise truck smaller than an American Ford F-150 pickup truck. After gathering cargo from different small factories, some guys at a warehouse consolidated our cargo onto a larger box truck. Then we rode on the highway toward Shanghai, which was a perilous adventure. The highway was filled with every kind of vehicle imaginable, including double semi-tractor-trailers, single tractor-trailers, big box trucks, little box trucks, little pickup trucks which were mostly overloaded with chicken coops and live animals of assorted kinds, tuk-tuks, rickshaws, mopeds, and bicycle carts. There was no average speed. Each vehicle was moving at its own pace and dodging the others as best it could. Since there was no overpass or crosswalk, pedestrians regularly ran across the highway. We passed a police car, which was dealing with an unfortunate pedestrian on the ground; he had just been mowed down by a large semi-tractor-trailer truck.

After we arrived at the assembly plant in Shanghai, I was ready to trace the last major step in the chain before the phones were shipped overseas. An independent logistics company managed the inbound and the outbound, the parts and the finished goods, the schedules and the surprises. My old friend Dan Middleman worked high up in KOL Logistics, and had for the last few years been stationed in Shanghai. KOL was originally called King Orient Lines, which was an American container shipping line until it was acquired by foreign interests. After the foreign acquisition, the company became much more than a liner shipper operator. It became a full-scale global logistics management company that happened to own some ships.

Dan managed marketing for KOL Logistics. He was responsible for developing total logistics solutions from end to end. His job was similar to Rob's in the Mexico-US trade, but Dan worked on the Asia Pacific trade route and handled any product that would travel overseas in a container. I caught him in his office in Shanghai on my way back from the factory tour. Dan, who had brown hair that was beginning to gray with wisdom, was among the best marketing and sales brains in the shipping industry—a quick and also a conceptual thinker. He developed new markets based on his keen insights into how to reduce cost and lead time for shippers in an environment where the market pressures were driving them the other way. Dan welcomed me into his office and we chatted about the smartphone supply chain. As I asked a few detailed questions, he asked his account representative for smartphones to join our meeting.

A young Chinese man named Patrick Jiào handled the smartphone account. Patrick was thirty-five years old and had a chiseled face that looked more like the face of a rock cliff than the soft, curvy facial features of many Chinese people. He was tall and thin and he stood up straight, but his voice, a tenor, belied the strength and stability that flowed from his outward appearance.

Patrick's job was to make sure that all the component parts got from the factories where they were made to the assembly plant, and that the finished product got out the other side of the factory and into the trucks on schedule. He kept a weekly production schedule in his back pocket, and checked with the various factories about their actual production every couple of hours. He spent a great deal of time troubleshooting delays and problems, and expedited late shipments in order to have enough parts to put together the finished products. He also managed the warehousing of the finished products, and was responsible for making sure there was enough inventory on hand to meet the export shipment schedules.

Patrick's father had been a professor, and Patrick liked to solve problems. He spoke English with hardly any accent, but when working in English he sometimes said things that were unrelated to the topic being discussed, leaving the other person wondering if there was a previously undetected language barrier. Regardless, business was booming, smartphones needed to be shipped, and he had the right skill set, at least on paper—inventory and logistics, supply chain management, and English language.

Patrick's inventory management responsibilities became critical when there were hazardous goods involved, since they could be responsible for toxic waste, including some that is radioactive and never completely decays. Many electronics components contained resins and chemical compounds of trace amounts of radioactive material, which if disposed of without care could have caused harm and even genetic mutations to natural flora and fauna. Most of these mutations occurred over long periods of time so the effects were usually not visible to the general public. However, there were some tragic incidents of dangerous warehousing practices, and Patrick tried hard to make sure his product was never in the headlines for the wrong reasons, like a Tianjin (China) explosion that killed 170 people. The logistics company that operated the warehouse had been illegally operating in the facility for two years. The facility itself was located adjacent to public buildings and facilities, in violation of a zoning statute. The authorities had not audited the records so there was no public record of the fact that compressed air, flammable and corrosive substances, oxidizing agents, and toxic chemicals were being stored in the warehouse.[4]

Part of Patrick's job was to manage the transportation chain. Many of the parts had to be trucked in from inland villages and cities hours and even days away, to Shanghai. The heavy truck traffic caused a lot of air pollution. China was only on Vehicle Standard 3, while the United States was on Vehicle Standard 6 and Europe was on Vehicle Standard 7, and because there were no clean fuel standards. China and Russia (which is more Asian than European in its vehicle emissions policies) were at least five years behind the United States and Canada. "China 5," which was scheduled to come into effect in 2018, still would not put China where the United States was in 2010. South Korea and Japan were exceptions—they were on the equivalent of the US Tier 2 Standard (Euro 6) in 2010.[5]

In addition to automotive air pollution, China relied very heavily on burning coal to generate power, and this also produced air pollution.[6] Until 1999, China produced more than five times as much coal than oil or gas. Air pollution in Shanghai was unacceptably high—not as bad as Beijing, which had a public health crisis over it, but not healthy.

The lack of planning and zoning made truck transport through the city challenging. While London and other major old-style cities emerged over decades, even centuries, in an organized way, Ho Chi Minh City, Shenzhen, and inland Chinese cities such as Xiaoshan were emerging chaotically, with urbanity usurping agricultural land in an unorganized way, ambiguity around land rights, and occasional random retrofitting of infrastructure such as roads, water, and electricity. The result was inconsistent infrastructure, crazy traffic jams, and huge variations in lifestyle, standard of living, and housing prices from house to house and neighborhood to neighborhood.

The finished product was exported from Shanghai, where it was assembled. Most electronics products were exported by air freight because inventory was expensive to hold, and also because the moisture involved in ocean shipping could ruin the internals of the products. Back at the airport, I noticed that letters didn't line up right on signs, and some letters were bigger than others. The signs were obviously handmade. This was very unlike Malaysia, where the quality standards seemed much more Western.

As I made my way through the airport toward the cargo area, I noticed that the employees of the airport were all dressed in police uniforms—thousands of them. The uniforms were formal and drab and didn't fit well, and men and women were wearing the same utilitarian uniforms. It felt like a police state. The people working in the airport export zone wore worn, wrinkled-faces, and wore cheap clothes. China was growing rapidly, but middle-class jobs were new to China.

The export procedures were insanely bureaucratic. The procedures for passenger traffic illustrated the point. Passengers had to fill out an arrival and a departure card, not just an arrival card, and they really checked paperwork both inbound and outbound. You had to go through customs and retrieve your luggage, even if you were just connecting to another flight. Then you had to go through customs again and security again before you could board your connecting flight. For one connection, it took me two hours of bureaucracy in the airport, and all I did was go back to the same gate I had come in at!

From Hong Kong, most of the smartphones were shipped to the United States. Retailers like Walmart and others in the United States ordered containerloads of the equipment, and distributed it through their own distribution centers and those of third party logistics companies (like KOL Logistics) to their stores. I knew a senior vice president of MarkIt-Down from some benchmarking I did a year beforehand, so I called to ask him about how he views the supply chain for smartphones.

Jay was a serious, numbers-driven sixty-one-year-old guy with thinning gray hair. He was a good communicator, and always liked to be in control. As long as things were in control, he was a happy man and smiled a lot and was very friendly to be with. When things got out of control, his authority came down like a sledgehammer. Jay believed firmly in the ideals and vision of Walmart. While some people engaged in a raging debate about whether Walmart was good or bad for the United States—helping to raise living standards for the average shopper, or sucking away jobs and giving to the Chinese, offshore—Jay did not have a doubt in his mind. He was "all-in" on the vision of "hard discounting," and wanted to make the company as large and profitable as possible.

Jay, and more specifically the people under him, had the responsibility of making sure that the incoming product arrived on time to meet customer demand without stock-outs. Their big challenge was the length of the supply chain between all the suppliers in Southeast Asia, the assembly operation in China, and the stores in the United States. It took four to six months for an order placed in the United States to arrive from China. Many steps needed to take place from placement of the order to manufacturing of the product, and then the ocean transport took a month. Plus, customs clearance and port operations took two weeks on each side of the ocean. As a result, inventory was frequently out of balance—both shortages, and later overstocks of the same product, all due to missed delivery dates.

I knew from working with other American consumer packaged goods companies like PenCo, which sold their products to retailers like MarkIt-Down, that their supply chains were often a mess of zigzagging across the globe. Outsourcing to China had grown too fast for most companies to formulate a logical plan and transition in an orderly way. They tried to graft "the China thing" onto their previous soft and hard infrastructure (people, skills, and factories in the United States, for example) rather than do a massive but efficient, even optimal, reconfiguration of their extended supply chains. They also didn't have enough experience with China yet to formalize and lock in long-term commitments. Not enough people in their ranks had been to China, knew enough about where to source from, or spoke Chinese. There was an enormous learning curve. Therefore, most companies did a lot of shipping of components to and from places in China and Asia, to the United States and back to China for assembly, and back to the United States again for sale. This was economically possible because there was no carbon tax, and gasoline was cheap.

As I wrapped up my trip in Shenzhen, I asked Alex to do his best to compute the D-Factor for the smartphone based on what I had been able to tell him about the supply chain and operations for the production and distribution of these wonderful, life-changing devices. He said that the total D-Factor was 630—the equivalent of drinking 350 bottles of water. Mining of the supporting materials contributed directly to the depletionary impact (*Depletion factor of 258*) of this supply chain. Silica, bauxite, crude oil for the plastic components, rare earth oxides used in the battery and electronics, and wood products for the packaging, all depleted natural resources, albeit on a smaller scale than fuel production. The logistical movement of the finished products took place by air freight due to the high value of the inventory, and to prevent rust and water damage from ocean containerization. Air travel contributed almost twice the D-Factor as the device itself. Leaching in the mines and air pollution caused by mining activities contributed to a *Devastation factor of 272*. In addition, the undisclosed toxic wastes from the display and processor made in South Korea contributed to the D-Factor. Subcontractors were not under the same environmental regimes, so they cut costs by disregarding environmental factors. Unmonitored work conditions in the myriad small subcontractors in India, Korea, and China contributed to a *Dehumanization factor of 101*. A wide variation in working conditions, working hours, vacation and sick day and family leave policies, and compensation contributed to a small proportion of workers that were stressed, emotionally or harassed, and some of whom have no viable career path or alternative job options.

If American consumers had bought their electronic devices from European or American manufacturers, for example from Nokia (Finland) or Hewlett Packard (US) instead of Apple products made in China or Samsung products made in South Korea, the D-Factor would have decreased by 18 percent, as stricter environmental regulations in Europe and the United States would have regulated the disposal of chemical by-products from microchips (silica resins), batteries (lithium oxide), and plastic injection molding manufacturing (polystyrene) that was relatively unchecked in China, South Korea, and Taiwan. Cleaner air standards would have required more efficient production processes, and assured a more sustainable and healthy environment for workers and the community surrounding the manufacturing operations.

However, the cost would have increased by 11 percent. Would consumers have paid it? The industry was hotly price-competitive, and an 11 percent price disadvantage would definitely have cost market share.

Although I was examining the supply chain for exporting electronics from Asia to the United States, growth and consumption within Asia itself was on its way to contributing more negative externalities than the export trade did. Rapid increases in construction and consumption in developing Asia over the forty previous years drove much of the increase in industrial production. The largest single cause of the rise of industrialization and the consequent side-effects was the population growth in Asia itself.

China consistently attempted to manage the adaptation to the industrial boom that has occurred, but as of 2010 it was hard to keep ahead of the rapid growth. Asian countries' greenhouse gas (GHG) emissions increased 1,600 percent, from 0.42 to 7.06 Gt, between 1990 and 2010,[7] compared to 31 percent for OECD countries, which increased from 3.26 to 5.26 $GtCO_2$.[8] Asia's GHG emission from industry grew from 15 percent of the world's total in 1970 to 52 percent of the world's total in 2010, and accounted for virtually all of the increase in industrial GHG emissions during that period. China accounted for the single largest contributor to GHG emissions in the world in 2010.[9]

As I pondered the potential solutions, my phone rang. Looking at the screen, I saw a picture of my daughter Camille. I hadn't seen Camille since I started this trip twenty-six days ago. "Hi, when are you coming home?" she cuts straight to the point. The good thing about traveling eastbound is gaining time in transit due to the time zone change. It was Wednesday, and even with twenty hours in transit I can still be back on Wednesday by crossing the international dateline! "I'm leaving right now and will be there today itself! See you soon!"

NOTES

1. UNEP (2016), "UNEP 2016 Asia Environmental Enforcement Award," http://tropicalbiology.org/unep-2016-asia-environmental-enforcement-award/. Accessed June 15, 2017.
2. OECD (2006), "Water: The Experience in OECD Countries," *Environmental Performance Reviews*, OECD Publishing, Paris.
3. UN-Water (2016), "Water Pollution," *Environmental Degradation and Disasters* (database), www.unwater.org/statistics_pollu.html. Accessed in 2016.
4. Wikipedia (2015), "Tianjin Explosion," https://en.wikipedia.org/wiki/2015_Tianjin_explosions. Accessed June 15, 2017.

5. Untitled presentation by the International Council on Clean Transportation, published at TransportPolicy.Net. http://www.theicct.org/sites/default/files/Blumberg_11July2014.pdf, p. 1. Accessed June 14, 2017.

6. David Jacoby (2016), "Global Oil and Gas Demand Outlook and Supply Trends That Will Impact Upcoming Projects." CSSOPE, Beijing, May 2016.

7. IPCC (2014), *Climate Change 2014: Mitigation of Climate Change. Contribution of Working Group III to the Fifth Assessment Report of the Intergovernmental Panel on Climate Change*, Cambridge University Press, United Kingdom and New York, NY, USA.

8. Giga-tonne.

9. IPCC (2014), *Climate Change 2014: Mitigation of Climate Change. Contribution of Working Group III to the Fifth Assessment Report of the Intergovernmental Panel on Climate Change*, Cambridge University Press, United Kingdom and New York, NY, USA, p. 131. Followed by the United States (17.2 percent), India (6.6 percent), and Russia (5.1 percent).

TWELVE

Kuala Lumpur to Beijing

BEIJING, MAY 28, 2030

Tina and I arrive an hour and a half before our expected meeting with Feng Xie, a wealthy executive of the largest shipping companies in China, in order to be sure to catch him and so we would not cause him inconvenience by being late. It is Feng, in fact, who is late. Later, we manage to catch him downstairs at Club 16, a posh, private, and very exclusive club with dark mahogany paneling, marble bathrooms, and a smartly dressed maître d' whose job seems to be keeping nonmembers out. A faint smell of cigar smoke permeates the air, and he himself smokes a stogie as he plays Texas Hold'em. This is the oil-rich executive district for the burgeoning class of nouveaux riches in Beijing. Feng has become rich on deals he has cut in the shipping business. He has black hair and a medium build, and sports casual designer clothes. When we arrive, he continues playing his game, and even as I talk he seems uninterested in what I have to say. He does not have much incentive to listen; he has everything he needs.

The emergence of China's new class of ultrawealthy businesspeople is especially evident on Jinbao Street, a revitalized green area between the Forbidden City and the Central Business District. Beijing's "Park Avenue" or "Champs Elysees" is stacked with five-star hotels, chic restaurants, and designer boutiques. Exotic car showrooms line each side: Ferrari, Rolls-Royce, Maserati, Jaguar, BMW, Lamborghini, Aston Martin, and Mercedes. Most shopping malls in Beijing have gone upscale. Italian and French luxury brands have deeply penetrated the retail scene there. When I was a child, my aunt, who incessantly traveled internationally, used to bring me back small artisanal gifts from her trips to China. The country has come so far! When General Motors tried to rebrand one of its car models in 1989, it used the slogan "Not Your Father's Oldsmobile." Today's China is certainly "Not Your Father's China."

I have multiple goals for this visit to Asia: to attend a large oil and gas trade show in Beijing, to meet prospective clients in Beijing, to audit the factory of a valve supplier in Taipei for a Saudi Arabian client, and to understand the supply chain for smartphones to update my knowledge from when I visited Shenzhen in 2010—I am getting very close to finishing this "side project" that I started at the beginning of this trip. There are many things to accomplish on this trip because when I travel to Asia I usually have multiple goals to make good use of the cost and time involved in the trans-Pacific journey.

My company's agent in China, Tina Yang, meets me at the airport in Beijing. Tina has a long face, long black hair, and a broad smile. She wears jeans with a wide belt, and a man's shirt. She looks successful but comfortable. She started her own engineering staffing company eight years ago based on a staffing request from one of the big oil service companies, and has been headhunting for oil companies and their suppliers ever since. Her company places groups of engineers on projects abroad. She interrupts our conversation frequently to answer phone calls about a staffing project in Iraq involving Chinese expatriate engineers. It's obvious why she is successful at this business—she's assertive, she's an aggressive networker, and she's driven, energetic, and clever. She has a sidekick named Binbin, who describes herself as "a good girl." She is sweet—doesn't drink alcohol and won't go near Mahjong. She says she is thirty-six, but she looks twenty-six.

Tina and I spend the first day at the China International Petroleum Exhibition in Beijing mostly independent of one another, doing our own business. We meet for dinner and drinks, and she introduces me to Garry Kirkland, an oil tycoon who helps Chinese companies understand and exploit the full potential of wells. Garry has been involved in oil and gas operations for over thirty years, and participated in one of the most successful oilfield acquisitions in the United States. He and his partners invested over $200 million in the company's operations, and it paid off. He has chosen to live in Beijing for the last twenty years. In addition to introducing me to Garry that night, Tina also introduced me to Chinese rice wine (like Japanese sake).

On the second day we make a sales call to Breeze, a windfarm developer and wind turbine generator manufacturer on the outskirts of Beijing. There are five of us in the meeting. Edgar, an American, is an ex-consultant with McKinsey, and was also formerly with ITSys, the Indian IT giant. In addition, he has worked for Breeze's major competitor. Edgar runs the International division, and will be moving back to Los Angeles soon. Wei was an MIT grad who previously worked for NASA, the American space agency, and SKF, a multinational manufacturer of bearings and seals. He runs R&D.

These people are smart and educated, the best in the industry. The concept and business strategy of being a developer as well as a manufacturer of the primary component, the wind turbine generator, is a smart strategy. They aim to develop enough windfarms around the world to achieve economies of scale in production of the turbine. To this end, they are targeting onshore windfarms instead of offshore windfarms because the market involves many more units, even though the offshore market is potentially very lucrative because of the high cost of initial capital expenditure. We talk about the Cape Wind project in the United States.

Edgar and Wei clearly regard themselves as the world experts on wind turbines, bearings, and windfarm development, and they both tell me and Tina that there is nothing we can tell him that they don't already know. They do not ask what we have to offer; they are not even interested in a pitch. In fact, they themselves want to sell their research services to other windfarm operators. We don't even have a chance to make a case for any of the proposals we have prepared. I think about how dramatically the situation changed over the last ten years. Previously, Chinese companies like this were in dire need of the knowledge and technical expertise of companies like mine. The problem in the past was that they didn't have the money to pay for our services. Today they can pay for us but don't need us. Now the United States seems like the developing country and China appears to be the superpower.

The third reason I was in China on this trip is to audit a Taiwanese valve supplier for a Saudi Arabian client that served the Saudi oil and gas giant ArabaCo. Ten years ago, ArabaCo and none of its immediate suppliers would have considered using Chinese or Taiwanese suppliers, especially for something as important as valves, and this was for good reasons.

To begin with, Chinese suppliers almost universally had very weak English language skills, particularly among the top management, leading to a situation where one or two junior staff were in almost complete control of the relationship. Whereas in other countries, the top management spoke English and the lower level staff did not, it was the reverse in China. In China, senior management was not always well educated, and even if they were, they didn't necessarily learn English. The Chinese market was insular for most of their lifetimes, and most, if not all, of their business was conducted in Chinese. In contrast, the junior staff learned English in school. Therefore, it was much more likely for junior and young people to speak English than the senior managers and executives.

Furthermore, due to cultural differences, the junior salespeople were clearly "selling" the product; they rarely had engineering or technical backgrounds or engaged a consultative selling process that is de rigueur in the West for complex engineered products like industrial pumps and compressors. To a buyer from the West or the Middle East, it was unsatisfactory to have a relationship with one or two junior staff in their twenties. Sometimes they passed on messages to their senior managers, and less commonly they directly translated some comments from the managers, often they just controlled the relationship, so there was little to no real senior management involvement on the Chinese side. In this connection, it did not help that the junior Chinese sales staff who, for reasons mentioned above, were perceived as incompetent by Western managers, were mostly female, since this reinforced a stereotype in the minds of some buyers.

The language differences led to another odd and equally unsatisfactory aspect of the relationship between Western buyers and Chinese sellers. Because Chinese names were so hard to pronounce for Westerners, the junior, young Chinese salespeople almost universally adopted Westernized pseudonyms. These were usually very simple one- or two-syllable names, like Sally or Sara or Tina for young ladies and Andy or Mike or Tony for young men. I understood the rationale for this, and usually accepted it even though I felt that the nature of the relationship was somehow obfuscated by use of a pseudonym. My trust was shaken, however, when one of the salespeople I met changed her American name each time she changed employers. When she worked for the first company, her name was Tina, when she worked for the second company, her name was Mary, and when she worked for the third company, her name was Linda. In her mind, I guess this helped keep things clear—I suppose she was able to pretend that she didn't have any conflicts of interest between her work at the different companies. However, she explicitly approached me as a customer to try to convert me from her old company to her new company, so the name-changing protocol obviously didn't avoid conflicts of interest.

Also, the vast majority of these suppliers were too small to have enough product sophistication, skills, or manufacturing capacity to supply most Western industrial companies. Trade shows were often divided by country, with each country having a pavilion. I went to dozens of trade shows where the Chinese pavilion consisted of hundreds of small, shared microbooths, and sometimes one sales representative covered multiple companies. Based on information I scoured beforehand, and my strategy for covering all the suppliers I needed to meet at the tradeshow, I identified specific companies of interest. For the Chinese suppliers, I often had trouble locating them, then when I did find them there was very frequently a solid language wall between us, with neither of us speaking a word of the other's language. This bode very poorly for any potential relationship between my clients, which were large industrial companies with safety-critical installations, and these small Chinese shops were not the appropriate partners for that.

In 2010, most of the Chinese suppliers, especially the small ones, had no industry focus. As a result, they frequently only produced one or two of the full range of oil and gas pumps my clients were looking for, and the applications where they had installed these pumps varied from food and beverage to healthcare to automotive, and, maybe one oil and gas company.

Finally, their customer base rarely included companies in Europe or the United States. They sold their products wherever they could, which often meant countries such as Congo, Kazakhstan, Bangladesh, and Iran. These countries didn't have much money, so needed to buy at the lowest price, which the Chinese were great at offering, or were embargoed by the United States, the EU, or in prior years the UN, so had a limited choice of suppliers.

The Taiwanese suppliers were often more suitable as suppliers to Western manufacturers than the Chinese ones, but Western buyers typically lumped them into a single category, which was not good for the Taiwanese.

I am about to find out that both Chinese and Taiwanese manufacturing have changed dramatically over the past decade. I fly into Taipei and take a taxi downtown. My hotel room has a stellar view of the harbor; I can surveil all the shipping activities.

Andrew Wang, general manager of RK Valves, picks me up from my hotel at 8:30 AM. He has a broad face with a healthy suntan from a recent holiday. He is wearing a polo shirt and jeans—comfortable, and not feeling the need to impress me, even though I am the auditor from a big Middle Eastern buyer. He wore a poker face during the meetings and tour, and even lunch. He was evidently thinking and evaluating our potential.

Andrew is the oldest of three brothers. Mo, the middle brother, is taller than Andrew. Respectful of his older brother, Mo stays quiet in meetings, never interrupting and never disagreeing with Andrew. Outside of meetings and the office, Mo is jovial, jocular, and extroverted, and he communicates more openly with his brother about his opinions, but would always defer on important matters and in any area of potential disagreement. Andrew manages the factory operations, finance, and international operations. Mo, the middle brother, handles domestic accounts, and Yu, the youngest, handles inventory and purchasing.

KR Valves was formerly government-run and was privatized twenty years ago. Andrew and his two brothers invested in it during the initial public offering, and have progressively increased their shares and their management responsibilities. Andrew, the oldest of three brothers, manages the factory operations, finance, and international operations. Mo, the middle brother, handles domestic accounts, and Yu, the youngest handles inventory and purchasing. Andrew, with whom I have the most interaction throughout the day, stays close to people he knows. He is not an extrovert; he makes no pretenses, and may seem to lack social skills. However, he is an entrepreneurial engineer and a savvy and successful businessman.

Andrew and his brothers and the extended management team have transformed RK Valves into a private sector market leader. Today it has a 330,000 square meter workshop with 300 technicians who all have college education. The factory produces 50,000 units per year. The production process is based on 3D printing, and one work cell consisted of a very expensive 3D printer that produces each unit to custom specifications. The workers are a mix of Japanese and Chinese, and are well-educated and well-trained engineers with experience in oil and gas valves. The company is a joint venture with a Japanese company, and exports are shipped to a distribution center in Japan, where complete orders are fulfilled.

The company is a vertically integrated ISO 9001 facility having everything from foundry to finished goods all in a relatively compact footprint. It is almost at the level of the top European competitor, which competes in the premium, custom market space. It has an impressive array of measurement, calibration, and testing facilities, as well as the 3D printer. Everybody on staff speaks English fluently and easily. The company sells extensively to Chinese government utilities and process manufacturing firms, and has grown its export business from 2 percent when they took over the company, to 35 percent today. Ten years ago they sold entirely through traders, but today 80 percent of what they sell is through direct relationships with customers, including the foreign ones. Revenues jumped 20 percent last year when they completed part of a mega-project to enhance infrastructure for the Trans-Asian oil and gas pipeline network.

My audit is complete, so I return to Shanghai.

From the hotel in the middle Chaoyang district, I walk around the block to the China World Mall. Strolling around, I walk past Bottega Veneta, Fendi, Hermes, Givenchy, Cartier, Canali, Louis Vuitton, Prada, and Versace, just to name a few high-end luxury brand stores. This is just one of at least five malls within a few minutes' walk, all of which contain the same dazzling assortment of upscale stores. These luxury brands have the same penetration in China as Starbucks has in America—on every streetcorner. If I was desperate to dump tens of thousands of dollars quickly on diamond necklaces, exotic scarves, or any other luxury goods, I could keep going to three other malls each within a few minutes' walk, including the Park Life Shopping Center, the Kerry Center, and The Place.

China has cemented its position as a global leader of renewable energy through its low-cost mass production of solar photovoltaic (PV) modules. Even in 2017, China's growth in solar was the largest (in megawatts [MW]) in the world, and outstripped any other market. Back then, the writing was on the wall. China's growth was almost double that of the United States and India, and more than triple that of Japan. All other markets were much smaller. Almost two-thirds of new PV installations were in Asia, even in 2017. Asia, mostly China, has continued for decades to consume more than half of all new solar installations and lead all other countries in installing solar capacity overseas.[1]

Since 2010, Chinese technology companies have developed English-language versions of Chinese killer apps that have replicated and overtaken American companies such as Uber, Dropbox, Google, and Facebook. While the Chinese apps were originally offered only in Chinese, they now have more English-language users worldwide than their American equivalents. The new market leaders are Didi Chuxing for ridesharing, Goodsync for file sharing, and WeChat for instant messaging. Google and WhatsApp remain strong in the West, but Uber is searching for a new strategy like SmartPhone was in 2015. Dropbox went into server hardware and networking, thinking that vertical integration would allow it to earn higher margins than virtual storage alone.

The newfound Chinese superiority is not limited to software. Chinese hardware has become ubiquitous around the world, partly buried inside the boxes of Western brands and partly under its own labels, thanks to compatibility and networking protocols in China that favored Chinese equipment. The resulting better technical performance of the Chinese equipment inside the Chinese market effectively protected the Chinese market and allowed the Chinese manufacturers to achieve economies of scale beyond what the Western companies could hope for.

The broad economic growth has allowed China to make great leaps in sustainability and quality of life, beyond what would have been imagined by Western pundits in 2010. China converted all its coal-burning power plants to natural gas, imposed a carbon tax on fuel for transportation, and mandated 100 percent recyclability of consumer products.

While other countries now rely on an extensive scorecard of metrics to track their progress toward sustainability, China does not bother with that. The government simply issues broad mandates that simplify the path toward sustainability. The federal government offers limited consultation opportunities to the provincial governments, and provides little leeway for provinces or private sector companies to evade, avoid, or circumvent the policies. Penalties for noncompliance are severe. Therefore, progress toward the goals is assured and predictable, and does not need to be monitored, corrected, and adjusted subject to varying degrees of success in implementation. The approach is draconian, but it works.

REDISCOVERING SMART PHONES

Before embarking on the trip, one of my analysts identified the company and the person I should talk to in Shanghai to get an overall understanding of the supply chain for smartphones. The analyst set me up with a man named Lemon Li Fung. Through email and WhatsApp, Lemon and I agree to meet on the morning of my third day in Shanghai.

Lemon Li Fung picks me up at my hotel, and I begin tackling the next objective of this Asian trip: understanding how the supply chain for smartphones has changed since my visit to my friend Tony Tan in Shanghai in 2010. Lemon handles inbound and outbound transport and logistics for the outsourced electronics manufacturing company that makes smartphones in Shanghai. Lemon is young, about twenty-seven, and lanky. He got his degree in operations management three years before, and had one previous job managing inventory for the Chinese electric and gas utility. Lemon is extroverted, but has limited fluency in English. He is assertive and always trying to do better, grow, network, and rise professionally, so he works hard to speak to me in English. His English is broken and difficult to understand, but I appreciate his efforts. Lemon has an office in the airport complex and does most of his management on the computer. Eight inventory staff report to him on a dotted line basis and to the business unit manager on a solid line basis.

He explains the process to me. Component parts come in from throughout Southeast Asia, and a freight forwarder called GlobeTran consolidates the incoming shipments and delivers a steady flow of components to the assembly plant throughout every workday. Lemon does "paperwork" all day long, but he doesn't have a shred of paper on his desk. He works in various applications on his computer, all of which are enabled also on his mobile. Once he has configured the inventory parameters, he mostly manages by exception. Of course, the peaks and valleys of demand patterns at major US customers create a lot of exceptions at certain times. He also spends a lot of time preparing for the migration from the current to the next generation model. He will need to purge the old inventory and replace it with finished units and parts and components for the new model. I recognize the similarities between assembly of the smartphone in Shanghai and assembly of the smartphone in Shenzhen many years earlier.

We walk through the receiving area in the airport. Shanghai's airport is new and clean, and the flow of its cargo operations is efficient and high-volume. A logical physical layout helps move product along, and an extensive system of conveyors with automated sensors, shoots, and diverters makes the operation seem like the UPS hub in Louisville or the FedEx hub in Memphis. Although the operation is "just-in-time," GlobeTran operates a warehouse in Taipei in case of delays in transit.

As we walk, he stops for a coffee with the customs agent, Qiandy Mao, who goes by her Western homonym, Candy. She is short, thin, and young. Candy processes imports for the government. In the old world, she would have processed a lot of paper forms, but today she mostly uses apps to check for completeness of information, and routes exceptions to people in other departments for action. For example, if value information is missing, she notifies the tax department. If safety information is missing, she notifies the health and safety department. If shipper or consignee information is missing, she notifies the customer service department, and so on.

Customs clearance, which used to take almost two weeks in 2010, now takes less than two hours because of digitized, more accurate manifests, electronic preclearance at the ports of origin, and intelligent algorithms that identify cargo that is more likely to have anomalies and require inspection, and pass through all the rest of the cargo that has no customs or fiscal irregularities. As we continue our discussion, we step into her office. She takes a phone call and signals us to hold on for a minute while she resolves a discrepancy with the customer service department. As I observe her interactions with others, I notice that she is assertive, speaks her mind. I think: she would probably be very good at a sales role. We continue our walk through the airport, and Lemon drives me to the outsourced manufacturing company that assembles these component parts into the final product.

At the assembly plant outside of Shanghai, we meet Bill Ryan, a tall, blond American man in a sea of short Chinese people. He does not speak Chinese, but his managers spoke enough English for him to be effective. He begins by explaining some background on his company, Micro-Electronica. I knew about the company because it was disclosed in the brand-name manufacturers annual report. In fact, it has acquired a certain degree of fame. The public interest is significant, so the company developed a visitor tours like you find at BMW in Bavaria or Harley-Davidson in Milwaukee. It has an online reservation system for booking those tours.

Micro-Electronica is an electronics manufacturing company, and Bill's job is to coordinate the inbound and outbound flow of materials, components, and finished products. Since electronic products like the smartphone have hundreds and sometimes thousands of parts, and manufacturers are under tremendous cost and quality pressure, Bill's job is one of the most challenging links in the supply chain.

The smartphone is an upscale product, whose manufacturing has strict specifications for metals and all other raw materials. The use of scrap metal for some of the components has decreased, and where it happens, world-class environmental standards apply and are enforced. This has made it easier for China to eliminate the dumping of scrap metal into the country. In 2013, it strengthened laws on the importation of scrap, and began charging increased demurrage on ships waiting to unload at port, to limit this dumping,[2] and it has tightened the requirements several times since then.

Manufacturing has become more vertically integrated and less specialized by plant since China instituted a progressive carbon tax in line with UN Global standards. The vertical integration generated more local value-added, helped develop more integrative knowledge-based work, and reduced the amount of shipping and warehousing to one outbound flow from the manufacturing plant to the consumer. Huge logistics parks and mega ports were scaled back and their real estate was better integrated into the surrounding communities. The real estate that was conserved through the streamlining of logistics networks is now a combination of housing and mixed-use commercial developments.

Compared to 2010, the factory workers are more satisfied with their work (according to the latest human resource survey), and this was evident from our brief interactions with them on the plant floor. They have a more satisfying work life in general since the top-down autocratic command-and-control work environment has morphed into a more European-style collaboration between management and workers. Workers councils are now mandated by law, following the European standard, and whistleblower hotlines help to check employer abuses of power. The Labor Department now tracks key performance indicators countrywide. Work-related suicide rates have been all but eliminated, and employers identify mental illness issues in mandatory screens during the hiring process, rather than leaving undetected mental illnesses to result in workplace social incidents including violence.

Lemon continues to trace the supply chain by driving me along the same route that the truck takes from the warehouse in the south back to the airport, from which the finished product was exported around the world. Shanghai is more livable than Shenzhen was when I was there. There are parks and more community social activities. A newly formed Ministry of Culture helps to catalyze the integration of film, concerts, theater, and other creative events, and allocates budget for the venues and some compensation for organizers. The national dialogue about progress and development has evolved considerably, from near zero the last time we were here. The five-year planning process is online and organized and interactive crowdsourcing process that engages not only politicians, but also businesses and communities, in the development of regional and local plans.

We drive with the windows open. The air quality has improved since the last time we visited. The PM 2.5 rating for this day is 61, half of what it was when I was in Shenzhen before. As part of a convergence scheme whereby all industrial countries attained European air pollution standards within a twenty-year period, the Chinese auto fuel emissions standard has been tightened to 2010 European levels, all new vehicles are running on ethanol blend gasoline, and Chinese power plants have undergone a massive transformation, moving from coal-fired to 80 percent driven by natural gas the rest from renewable fuels like solar, wind, and biomass. There has also been a major shift from central power plants to distributed solar power generation—local production of electricity for each shopping mall, hospital, residential development, etc.—as huge reductions in the cost of solar technology has made this more economical in some cases than buying electricity from the grid. These local systems also supply power to the grid when they have surplus capacity, which reduces cost on the grid, too.

We wind through many small streets as Lemon is guided by his GPS system. The city has implemented a Smart Traffic system that uses twenty billion sensors, RFID, and Big Data mining and algorithms to measure, monitor, and optimize flow patterns in real time through traffic signaling and real-time data, which links to GIS maps and routing software in his smartphone.[3] The ride back to the airport is quick—only twenty minutes.

As with the inbound cargo, finished smartphones ready for export undergo precustoms clearance in Shanghai so that when they arrive into the importing countries they can proceed directly to the retail and warehouse without delays at the airport.

Once the product gets on the plane, the freight forwarder takes responsibility for it. The global freight forwarder is the same one that handled the imported products, GlobeTran. GlobeTran handles consolidation of exports in a facility adjacent to the import consolidation facility, an arrangement that provides economies of scale, consistency of processes and information systems between imports and exports, and shared overhead costs. A win-win deal for everybody, compared to the previous arrangement in which multiple small importing distributors in multiple small exporting distributors each operated parallel but slightly incompatible gathering, stocking, customs clearance, and handling, all at high cost and without coordination.

The forwarding operation is managed by Chinese nationals, but shipments to America are overseen by an American. Lemon knows the fellow, and to avoid struggling with English for the next twenty minutes, Lemon introduces me by speakerphone from his car. He dials and puts a man named Bob Gannett on the phone. He makes it a videocall for my benefit. Bob appears to be a distinguished gentleman with chiseled features and wavy silver hair. He has spent his career in electronics logistics: previously he worked for two other freight forwarders. He lives in Boston and visits China and Taiwan every eight weeks to oversee GlobeTran's operations there.

I inquire about shipments to Bogotá, since that is where I started my journey tracing supply chains two decades ago. Bob explains that EVA Air, or any other one of a number of prequalified carriers that offers the right schedule and price, is selected by dynamic auction, carries the finished products away to Bogotá, which serves as the distribution hub for the Andean region of Latin America. Because of the high value of full containerloads of smartphones, security is tight in and around the transport operations.

He offers me contact information for the ultimate customers in Colombia and the United States, in case I want to continue tracing the path of the supply chain, and I took down some names.

By this point, I have enough information to give to Alex so he can update the D-Factor calculation. I send him my notes and debrief him by phone that night in the hotel (it was nine hours later where he is in Portland, Oregon).

His report is in my email when I wake up the next morning. The supply chain for smartphones is much more environmentally friendly than the supply chain for Blackberries was in 2010. ASEAN cooperation on tightening regulations on leaching and toxic chemical deposits has lessened the environmental damage created in the mining for raw materials like silicon and bauxite. At the same time, a harmonized system of engineering and design standards triggered the use of substitute materials for radioactive rare earths and other hazardous materials used as inputs from microchips and electronic components. Environmental conditions in China have improved dramatically, thanks especially to the carbon tax and the movement to clean fuels for power generation. The Depletion factor has fallen by 33 percent, a benefit which was especially felt in China and Australia, both of which have substantial mining operations that provide inputs to electronic goods. Tighter air pollution standards, emissions vehicle requirements, and intelligent transportation networks helped to reduce the Devastation factor by 33 percent. The working conditions for factory workers in China, and for ship crews and truck drivers in China and the United States improved considerably due to China's adoption of the UN Compact principles, and rising health, safety, and environmental standards from the Occupational Safety and Health Administration (OSHA) and other government agencies on both sides of the Pacific. In addition, adherence to the UN Global Compact and the promulgation of humane working conditions, and the reduction of discrimination in the workplace, yielded a 38 percent reduction in the Dehumanization factor. These improvements conspired to reduce the overall D-Factor of a smartphone by 34 percent, compared to the D-Factor for a smartphone in 2010.

The next day I start home from Shanghai to Boston. Reclining in my plane seat, I ponder the changes that have occurred over the last decade or so. Almost everywhere I have been on this round-the-world journey has become a better place! The air is cleaner, the water is cleaner, most jobs have acquired an element of technological skill and are more fulfilling. Supply chains and transportation networks are more efficient, and people from all cultures are more socially connected than ever before. How did that all happen? Was it a natural process of evolution, happening organically in millions of ways at the local level? How much of this was due to government regulation? Was there a coordinated international effort, or just each country doing its own thing? How much impact did technology make? I resolved to answer these questions.

NOTES

1. David Jacoby (2017), "Market Outlook for Solar PV." Presentation to SNEC Conference. Shanghai, China. April 20, 2017.
2. China Briefing (2016), "Trash or Treasure? Prospects for China's Recycling Industry," www.china-briefing.com/news/2016/03/10/trash-or-treasure-prospects-for-chinas-recycling-industry.html. Accessed June 15, 2017.
3. SAP (n.d.), "Smart Cities—SAP Helps Clear Traffic Jams," http://news.sap.com/sap-helps-clear-traffic-jams/. Accessed June 15, 2017.

THIRTEEN

The Road Back Home

BOSTON, JUNE 11, 2030

I spend several days at the office with Alex mind-mapping and outlining and comparing my two parallel trips two decades apart from one another.

After studying my notes, outlines, and diagrams, Alex and I conclude that the main driver of change over the past decade was not voluntary behavior change. Most reductions in D-Factors involved "hard choices" that consumers and businesses would have been unlikely to make without financial incentives or regulatory mandates. As a result of global supply chains, it is hard to disassociate consumption from Depletion, Devastation, and Dehumanization. The D-Factor *per dollar* of the examples in chapter 2 is consistently between one and three times the expenditure in dollars. In other words, sustainability is negatively correlated with the kind of economy and growth that we considered to be good in 2010; i.e., our old growth paradigm was unsustainable.

In 2010, local decisions were made on the most economical basis, which was usually not the most sustainable way. Many authors of the period claimed that sustainable practices were actually also the cheapest way to do things, because efficiency was cost-effective. The extension of this thinking was that the problem would go away on its own if we just kept doing the same things better. This thinking was idealistic and, given the large gap between the current state and the sustainable state, it was wrong. Major, not minor, behavior change was required to close the gap. In developed countries, most individuals still only made socially or environmentally friendly purchasing choices if the cost was the same as the current price. In emerging economies, hardly anybody could afford to pay more than the current, unsustainable, price.

Backyard burning fueled by burning automobile tires and scrap wood that emit toxic offgases was cheaper and more convenient than burning using cleaner fuels.[1] Driving a polluting moto was cheaper than the cost of getting it fixed or buying a newer, more efficient one. Solar panels required an investment, and consumers only made that investment if it was subsidized. If it was not subsidized, they continued to use fossil fuels. Electric vehicles cost more than gasoline-powered vehicles for the same grade of car, and consumers generally opted for the less expensive car. Incandescent lamps were cheaper than compact fluorescent lamps (CFL), so people chose to buy the cheaper ones as long as they remained available. It was cheaper to dump used electronics in the trash rather than pay the recycling fee. The list went on and on.

Only about 20 percent of the largest firms, and 5 percent of all companies, voluntarily implemented "labor human rights" policies related to their supply chains. European firms were more prone to doing so (30–40 percent), while Asian firms almost never enacted such policies on their own—only 10–20 percent of Asian companies had labor policies in place.[2] BearingPoint reached similar conclusions; it found that 54 percent of companies with sales over $1 billion considered supply chain sustainability, compared to only 29 percent of smaller companies.[3]

MANDATES

To get a clearer picture of how mandates occurred and the benefits they generated, I consult a European friend of mine. Europe has been at the forefront of sustainable capitalism for decades.

Hans lives in Munich, and I have had enough conversations with him over the years to appreciate his deep sensibility to social organization and goals within a capitalist context, and his sense of a larger sociocultural aesthetic and vision. Hans doesn't seem at all German to me. He speaks Spanish and has spent a lot of time in Venezuela and Mexico. Although he lives in Germany, he works in Switzerland and does projects in the Middle East. He has brown hair, brown eyes, and a medium build. He could blend into any country between Russia and Bolivia. He is a bit of a chameleon in other ways, too—adaptive, flexible, and multilingual. He also has a strong social conscience, and his concern for the world at large and the engagement of communities worldwide to make the right decisions on issues pertaining to globalization, growth, and development. He plays an extremely active role in his community, sitting on the board of directors of a group that monitors the development and zoning of his town. That group debates and decides on a wide range of zoning petitions, and makes decisions about how the town will grow. It debates the encroachment of commercialization and the unhappy side of globalization on small communities in suburban and rural Germany. I admire his energy and active involvement in such discussions at the local level.

Our conversation leads to a debate about the pros and cons of government mandates to effect changes in behavior for the collective good, as opposed to incentives (via taxes and subsidies), and encouragements (such as monitoring, measurement, initiatives by nongovernmental organizations, and grassroots efforts).

We conclude that mandates were extremely effective in achieving rapid and substantial organizational and societal behavior change, and that seven mandates in particular resulted in a substantially improved quality of life at the national, regional, and global levels. These were: (1) carbon taxes; (2) clean power mandates; (3) clean fuel mandates; (4) recyclability mandates; (5) mandatory reporting into the UN Global Compact; (6) minimum standards for human zoning, and (7) D-Factor labeling mandates.

Carbon taxes at the source had a direct and immediate benefit. These should be increased progressively over thirty years or more to allow fossil fuel intensive industries to refocus their business strategies and models to a new and different type of postindustrial growth. Examples of carbon taxation include the UK Climate Change Program, which was established in 2000 to meet both the country's Kyoto Protocol commitment of a 12.5 percent reduction in greenhouse gas (GHG) emissions by 2008–2012 relative to 1990 and the domestic goal of a 20 percent CO_2 emissions reduction relative to 1990 by 2010. A key element of the Climate Change Program was the Climate Change Levy, an energy tax applied to industry, commerce, agriculture, and the public sector. The levy does not apply to domestic customers, charities, or oil use and electricity produced through combined heat and power (cogeneration) units or with renewable sources.[4]

Strong clean energy mandates on the power industry catalyzed the shift from petroleum products to natural gas for power generation. As noted by the Intergovernmental Panel on Climate Change (IPCC) back in 2014, "GHG emissions from energy supply can be reduced significantly by replacing current world average coal-fired power plants with modern, highly efficient natural gas-fired combined-cycle power plants or combined heat and power plants, provided that natural gas is available and the fugitive emissions associated with extraction and supply are low or mitigated."[5] Many large power generators in 2010 had a phased percentage renewable energy ("portfolio") approach to fuel-switching, which needed to be formalized and mandated where it was not already. The clean power mandate opened the way for global air pollution standards and power plant efficiency standards. The American Clean Power Plan, which was enacted under US president Obama, mandated a 50 percent decline in fossil fuel use for power generating utilities by 2030 and a phased reduction of utility carbon emissions of 30 percent by 2030, versus a 2005 baseline. Power plant efficiency standards ensured that monopoly producers could not consume high amounts of carbon, even if they met GHG and air pollution standards.

Low-carbon fuel standards assured that the fuel that was being consumed was relatively clean. I drove a methanol-fueled car in Brazil in 1991. Subsidies for alternative fuel vehicles have continued to expand. Electric vehicle sales leaped with advances in energy capture and storage technology.

Mandates for 100 percent recyclable consumer products cut down dramatically on trash and landfill everywhere. A bold and aggressive tax incentive for a zero net environmental footprint over the product lifecycle generated pressure for retailers, consumer products goods companies, and also industrial companies to push back on their entire supply chain for reduced environmental footprint at every stage of production and distribution. Correspondingly, universal penalties for noncompliant waste disposal built on the success of RoHS. To police this, companies were required to report their environmental performance, including end-of-life plans for all products ("end-of-life" means what users were supposed to do with them after they were done with them other than throw them into the trash). Additional "devastation"-related regulations included: global minimum standards for the purity of air and water emissions; energy efficiency standards for capital equipment, such as IE4 motor standards; and subsidized energy audits and benchmarking studies.

Reporting into the UN Global Compact made working conditions transparent, which enabled rectifications and improvements. Reporting was essential to accountability, and accountability was critical to achieving the change that was required to address unacceptable D-Factors. EU, China, and US governments mandated corporate reporting for all publicly traded companies on their performance relative to the goals of the UN Global Compact. The UN Global Compact is an initiative renouncing ten dehumanizing practices. Reporting on a broad scale was the first step along the path to upgrading of conditions. Watchdog groups akin to Greenpeace and Human Rights Watch emerged to audit compliance and accuracy. The ten principles of the UN Global Compact are, in short: (1) Respect for human rights; (2) No forced labor; (3) No child labor; (4) No employment discrimination; (5) No restrictions on association or collective bargaining; (6) Adequate occupational safety (emergency preparedness, fire safety, industrial hygiene); (7) Precautionary approach to environmental impact; (8) Development of environmentally friendly technologies; (9) No corruption, bribery, and kickbacks, including gifts; and (10) Proper accounting and business records. As of 2015 to 2017, the Global Reporting Initiative (GRI) already provided a head start: 9,175 organizations had a profile in GRI's Sustainability Disclosure Database.[6] The largest and most prominent companies reported: 92 percent of the top 250 companies and 81 percent of S&P 500 companies reported on sustainability.[7]

Many cities made urban spaces more livable. There was more infrastructure for bicycles and walking, including pedestrian zones, and eventually self-contained communities or new towns, especially with balanced, mixed use, areas to facilitate community and minimize transport requirements. To provide more livable communities, working groups and colloquia formalized universal principles of zoning that supported balanced and natural urban growth.[8] These plans were based on maximum ratios of vehicle traffic: pedestrian density.[9] To provide a model, the American Institute of Architects published principles for livable communities, including: (1) Design on a Human Scale; (2) Provide Choices; (3) Encourage Mixed-Use Development; (4) Preserve Urban Centers; (5) Vary Transportation Options; (6) Build Vibrant Public Spaces; (7) Create a Neighborhood Identity; (8) Protect Environmental Resources; (9) Conserve Landscapes; and (10) Design Matters.[10] Public transport proved to be a necessary component of human zoning. In order to facilitate livable communities and support infrastructure for a livable industrialization, governments made commuting faster and less environmentally damaging by providing mass transit networks and urban planning for human-propelled transport like bikes ("urban land cover is projected to expand by 56–310 percent between 2000 and 2030"[11]). A global fund was necessary because the investment could often not be justified at the city level. Although thousands of cities had some climate change action plan, very few had the "cross-sectoral measures to reduce sprawl and promote transit-oriented development."[12] For example, Bogotá had no transit system and too much urban growth has already occurred for an investment to be profitable. A retroactive digging effort (a huge exercise in dislocation) would have cost too much and there were not enough riders to justify the massive cost. Amsterdam provided a model of human zoning. The city was, and still is, filled with bicyclists, who are reducing carbon emissions, staying healthy, and reducing congestion.

D-Factor labeling created an awareness of social and environmental externalities, a part of the buying process. The D-Factor has allowed consumers to know in which percentage a specific product was causing Depletion, Devastation, and Dehumanization, as well as the activity and the country where the externalities are produced. The EU took the first step in mandating D-Factor labels in 2020 for all products, and it extended the labeling to services progressively by 2025. Some advancements were required in order to account for value-added at each stage of production and in each country. These provided a new era of growth for Enterprise Resource Planning systems like SAP, which expanded its scope of resources to include natural resources.

SAP and most other information system solution providers added two new modules that can provide detailed breakdowns of resource Depletion, Devastation, and expanded their traditional human resources modules. These steps, as well as the reporting that they facilitated, were critical in helping to track and account for the use of natural resources (Depletion), the processing of air and water and its impact on the quality of those (Devastation), in addition to resources that were internal to each company (like machine hours). It also expanded the scope of its human resources module to track and ensure compliance to working conditions (Dehumanization). SAP's new NatResource1 module tracks carbon taxes due and paid, and GHG emitted. Its NatResource2 module tracks air and water quality; fuel cleanliness; emissions trading activity; solar, wind, and thermal power generation efficiency; and subsidies and tax credits for alternative fuel vehicles; and recyclability including actual end-of-life use. Its expanded human resources module tracks the availability to its employees of mass transit systems and infrastructure for bicycling and walking, both on and off its corporate property, and working conditions with detailed diagnostics and checklists to flag questionable working conditions.

The information that these systems provided the information they needed to generate their D-Factor labels. As with nutrition labels, these were subject to audit and could result in a product recall if the data was found to be materially inaccurate. The companies were also required to submit the underlying data to municipal and state governments, who used it to improve local conditions for employees.

CHANGING THE DEFINITION OF GROWTH AND THE MEANING OF PROGRESS

Since most of the world's business decisions in 2010 were based on stock prices and Gross Domestic Product (GDP), there was a lot of inertia pushing back against the idea of mandates. However, it became increasingly clear that Depletion of nonrenewable natural resources and overdevelopment was not only threatening, but actually tangibly decreasing the quality of life around the globe. An increasing number of national and global health crises such as toxic air in India and the flooding and consequent relocation of complete island and coastal populations, ultimately led the UN to step in to take a more direct role in measuring our progress toward sustainability.

I meet with one of my economic analysts, Adeep, to recap what happened. Adeep is an economist and an environmentalist who is specialized in environmental issues and renewable energy economics. I decided to challenge him with this question. I broach the issue over dinner with his soon-to-be wife in a corner of New Delhi called Gurgao Cyber City. Adeep is deeply passionate about the topic. Here is the gist of our discussion.

First, the UN and the EU both adopted the Genuine Progress Indicator (GPI) to begin to create an awareness of the distinction between growth and progress. The EU's early adoption of the GPI gave it a head of steam, as they say. He explained the calculation: the GPI subtracts from GDP negative activities such as war, oil spill cleanup, and hurricane relief, and adds to GDP positive activities that do not appear in any national accounts, such as child-rearing and volunteerism. The GPI is the most comprehensive and useful metric for measuring the externalities discussed in this book and engaging policy action that corrects them. The definition of GPI is intricate and has been developed extensively in practice. Here is a summary of some of its key components:[13]

Economically, GPI starts with GDP, but then it divides this by a measurement of income inequality called the Gini index and the Income Distribution Index (IDI), its relative change over time. It then deducts the cost of underemployment, including chronically unemployed, discouraged workers, involuntary part-time workers. It then adds or subtracts capital investment flows if they are net inward or subtracts them if they are net outward.

Environmentally, the GPI deducts a lot of environmental side effects that we don't quantify at all today, such as the cost of water pollution, air pollution, and noise pollution, and the loss of wetlands, farmland and soil quality, and forest. It also deducts damage caused by CO_2 emissions and ozone depletion, and the cost of replacing infrastructure that burns fossil fuels with renewable power and fuels. Socially, the GPI adds the Value of Housework, parenting, and volunteer work. Then it subtracts costs of social disruption such as divorce, crime, and excessive time devoted to work (it uses a baseline of 1969 for this calculation). Conversely, it adds the value generated by higher education and infrastructure such as highways and streets, while netting out the cost of commuting, which most people disdain and doesn't add any economic value. The GPI is a more practical alternative to GDP than other alternative progress measurement methods such as the Index of Sustainable Economic Welfare (ISEW), Net National Product, the Mean Happiness Indicator, the Human Development Index, and the Environmentally Adjusted Economic Performance Indicator.

Adeep fills me in on a number of other indicators that the UN considered but dropped. The "Mean Happiness Indicator" grabs my attention. The country of Bhutan—a landlocked Himalayan country bordering Tibet with a total population of less than a million people—conducted an interesting survey in 2007. It asked its citizens about their happiness based on questions related to education, psychological well-being, health, time use, cultural diversity and resilience, good governance, community vitality, ecological diversity and resilience, and standard of living.[14] Those metrics captured an important end result that did not correlate with traditional economic measures. However, they suffered from measurement bias across respondents and over time due to lack of an objective scale. The same was true of other, similar, attempts at measuring happiness, such as Gross National Happiness and Gross National Wellness.

The ISEW began to address many of the important externalities, but it was not as well-defined as the GPI. It started with consumption rather than production, and it deducted the cost of environmental externalities such as air, water, and noise pollution, and damage to natural resources, and the value of domestic services. It discounted defense spending. This approach made it more reflective of real value added than Gross National Product (GNP). However, it lacked specific methodological clarity on the quantification of social and environmental costs.[15]

Net National Product (NNP), which was defined as GNP minus depreciation,[16] failed to capture many important social and environmental externalities. On the positive side, both GNP and depreciation were known, and GNP was a proven traditional measure of international strength, which reflected national income as well as income earned by the nation from abroad less income earned by the foreigners in the domestic market.[17] Furthermore, applying depreciation presented a truer picture of sustainability. However, the metric didn't adjust for social or environmental externalities; it was distorted by prices and taxes, even at constant volume of production; and it excluded unreported services such as parenting and domestic work.

The Human Development Index focused only on two nonincome parameters—health and education—so missed many of the problems brought to light by the D-Factors. It combined three quantitative measures—health (based on life expectancy), education (based on adult literacy and school enrollment), and income (based on GDP per capita using purchasing power parity)—into one index, then divided the resulting index into quartiles.[18] The Human Development Index was used by the UN Development Program, and its methodology of indexation and quartiles was useful for comparing countries. However, since it was based on only three key metrics, it provided limited insight into well-being.

The Environmentally Adjusted Economic Performance Indicator (GDP per capita times a longevity factor and divided by GHG emissions) accounted for only two non-GDP factors. [19]

Adeep said that US and EU selection of the GPI over those other metrics, and their promotion and support for broad data gathering and reporting of the GPI, was especially helpful in catalyzing awareness of the negative impacts of industrialization and globalization, and consequently in stimulating action.

KEEPING AN EYE ON THE BALL

Defining a top-level metric like GPI is terrific, but I know from consulting to corporate management that effecting substantial change in large, complex organizations requires more than one metric. In order to know why the overall indicator went up or down, you have to track subordinate metrics. As I start searching for key performance indicators, I know I have to bounce my ideas off my friend Mike. Mike, who has consulted almost all his professional career since holding early positions at British manufacturing conglomerates, has a special talent for developing performance metrics and dashboards. He is originally from England. Always the gentleman, he defers to my thoughts first.

I lay out where we were in 2010, and where we are now in 2030, on a scorecard organized into three parts: Depletion, Devastation, and Dehumanization. For each of the three, he comes back with a series of graphs that show the evolution from 2010 to now. Due to the difficulty of assembling consistent data sets, he uses 2017 as the base year.

Mike's Depletion graphs track the following metrics from the baseline (2017) to the present (2030): Carbon Tax (2017 baseline was 1.5 percent/bbl, 3 percent/MMBtu global average, and current level is 18 percent/bbl,[20] 18 percent/MMBtu.[21]); GHG emissions per capita (the baseline was 6.4 tonnes, and the current level is 5.9 tonnes), Maximum Air Pollution Standards (the baseline was 58 μg/m3, and the current level is 29 μg/m3); Mandated Low-Carbon Fuel Standards (the baseline was thirty-four countries with mandates averaging 7 percent content of alternative fuel/renewable energy, and the current level is fifty countries with an average of 20 percent); Solar Photovoltaic Panel Generation Efficiency (the baseline was 17 percent and today it is 30 percent); Wind Power Generation Efficiency (the baseline was 30–40 percent, and it is currently 45 percent); Thermal Power Generation Efficiency (the baseline was 36.4 percent world average in 2014, which has risen to 42.8 percent), and Subsidies & Tax Credits for Plug-in Electric Vehicles (in 2017 subsidies and tax credits were offered in fourteen countries with an average of max $6,300 for the purchase of a plug-in electric car, and today they are universal).

His Devastation matrix has the following baselines and targets: 100 percent Recyclable Consumer Products (the baseline was 35 percent of products manufactured, and today it's 50 percent); Sustainability Reporting Including End-of-Life (the baseline was 20.3 percent of public companies and voluntary only, versus 100 percent of public companies and are compulsory today); Corporate Air Emissions Quality (the baseline was 92 μg/m3, and the current level is 64 μg/m3); Industrial Energy Efficiency Targets (the baseline was 20 percent of countries setting a target, versus 100 percent of countries doing so today); Industrial Energy Efficiency Financial, Fiscal, and Regulatory Measures (at the outset 70 percent of countries had such incentives, versus 100 percent of countries today);

Implementation of Minimum Energy Performance Standards on Electric Motors (the baseline was 65 percent of countries having them, versus all countries having them today); and Mandatory Energy Audits and/or Energy Consumption Reporting (only 25 percent of countries offered them before, and today all countries do).

His Dehumanization matrix shows the following metrics: Mass Transit System Availability (only 18 percent of transit trips were via rail, 90 percent of metro areas had mass transit systems, and 30 percent of metro areas had high-speed lines in operation, whereas today 36 percent of transit trips are via rail, 100 percent of metro areas have mass transit systems, and 60 percent have high-speed lines); Infrastructure for Bicycles and Walking, including Pedestrian Zones (originally municipalities spent only $0.90/person/year/percent mode share for cycling infrastructure and 18.5 percent of urban transit was nonmotorized, whereas today they spend $3.6/person/year/percent mode share for cycling infrastructure and 40 percent is nonmotorized); Industrial Slavery (the baseline was 0.50 percent of the population in slavery, versus none today); Global Subsidy Fund for Mass Transit Rail Systems (originally only 129 Metro and Light Rail projects were planned and underway, versus 400 today); and Mandatory Reporting of 10 UN Global Compact Conditions (originally only 18 percent of public companies reported them, versus 100 percent today).

BEING HONEST AND TRANSPARENT

How did companies (or, as academics say, "the private sector") adapt? For industrial policy issues, I knew who could give me a good run-down on which corporate reporting frameworks that had been instrumental in driving the improvement that I observed. I consulted my friend and expert Tony, an "old salt" who had worked on national policy development and regulation, including at the Carbon Trust in the United Kingdom. He was more "traveled" than anybody I know, especially in Africa and Asia, and usually in connection with some government policy work for lenders, nongovernmental organizations like the World Bank, or directly for governments.

It turns out that getting the EU and ASEAN to adopt the frameworks, methodologies, and analytic constructs of the Carbon Disclosure Project and the GRI was key to achieving common principles and metrics for the reporting of economic, environmental, and social performance. The GRI's metrics had already been endorsed by the UN Global Compact, and when the EU established a directive mandating that companies report on the GRI metrics, that cemented everything together. Independent auditors such as Bureau Veritas and DNV stepped in to ensure the accuracy of the figures. The aggregate statistics, which could be trusted because they had been validated, were used to inform policy decisions. Mandatory reporting into the UN Global Compact increased from 18 percent of public companies to 100 percent with the mandate.

EU directives were particularly influential in catalyzing change, especially the Directives on Environmental Impact Assessment and Accounts Modernisation and Transparency, which required environmental disclosures and key performance indicators.[22]

D-Factor labeling also played a part in raising awareness of insurers and investors, and by generating consumer awareness, which changed buying habits. Based on the work of a subcommittee of the UN Industrial Development Organization, more than one hundred countries mandated the use of D-Factor labels of one sort or another for every product they sold. For food products, the three key indexes for Depletion, Devastation, and Dehumanization, have most often been shortened to D1, D2, and D3, and appear on the nutrition labels that are widely used around the world. For other products and services, they follow a standard format. On the left, rows list the processing steps in the sequence that value was added, and across the top columns list the countries that contribute to the value added of that product or service. The total D-Factor is inside each cell of the matrix. It is conceptually the same as a calorie label.

NOTES

1. D. Vallero et al. (2012), "Society," *Unraveling Environmental Disasters,* Elsevier Science.
2. "Supply Chain Labour and Human Rights: S&P/ASX 200 and Global Company Policies." Australian Council of Super-Investors, 2011. http://www.law.harvard.edu/programs/lwp/pensions/publications/11_supply-chain_labour_and_human_rights_research_paper_-_final.pdf. Accessed June 15, 2017.
3. Inter-organisational Design of Voluntary Sustainability Initiatives: Increasing the Legitimacy of Sustainability Strategies for Supply Chains. Dissertation of the University of St. Gallen, Graduate School of Business Administration, Economics, Law and Social Sciences (HSG) to obtain the title of Doctor Oeconomiae submitted by Nils Peters from Germany. Approved on the application of Prof. Dr. Wolfgang Stölzle and Prof. Dr. Volker Hoffmann. Dissertation no. 3703, Gabler Verlag, Wiesbaden, 2010, p. 2.
4. L. Price et al. (2005), *Tax and Fiscal Policies for Promotion of Industrial Energy Efficiency: A Survey of International Experience,* Environmental Energy Technologies Division, The Netherlands, Lawrence Berkeley, p. 36.
5. IPCC (2014), *Climate Change 2014: Mitigation of Climate Change. Contribution of Working Group III to the Fifth Assessment Report of the Intergovernmental Panel on Climate Change,* Cambridge University Press, United Kingdom and New York, NY, USA, p. 21.
6. Global Reporting, "Sustainability Disclosure Database," http://database.globalreporting.org/. Accessed June 15, 2017.
7. "Currents of Change: The KPMG Survey of Corporate Responsibility Reporting 2015. https://home.kpmg.com/content/dam/kpmg/pdf/2015/12/KPMG-survey-of-CR-reporting-2015.pdf. Accessed June 15, 2017.
8. A more comprehensive set of principles is contained in IPCC (2014), *Climate Change 2014: Mitigation of Climate Change. Contribution of Working Group III to the Fifth*

Assessment Report of the Intergovernmental Panel on Climate Change, Cambridge University Press, United Kingdom and New York, NY, USA, p. 959.

9. 2009 World Development Indicators, The World Bank, http://web.worldbank.org/WBSITE/EXTERNAL/TOPICS/EXTTRANSPORT/EXTURBANTRANSPORT/0,,contentMDK:22345648~menuPK:8244758~pagePK:210058~piPK:210062~theSitePK:341449,00.html.

10. "What Makes a Community Livable? Livability 101," *The American Institute of Architects*, 2005.

11. IPCC (2014), *Climate Change 2014: Mitigation of Climate Change. Contribution of Working Group III to the Fifth Assessment Report of the Intergovernmental Panel on Climate Change*, Cambridge University Press, United Kingdom and New York, NY, USA, p. 25.

12. Ibid. Page 972 shows the GHG reduction targets for forty-two of them.

13. Wikipedia (n.d.), "Genuine progress indicator," https://en.wikipedia.org

14. Wikiprogress (n.d.), "Gross National Happiness," http://wikiprogress.org

15. Wikipedia (n.d.), "Index of Sustainable Economic Welfare," https://en.wikipedia.org

16. Wikipedia (n.d.), "Net national product," https://en.wikipedia.org

17. Difference Btw (n.d.), "Difference between GDP and GNP," http://www.differencebtw.com

18. Wikiprogress (n.d.), "Human Development Index," http://wikiprogress.org/articles/initiatives

19. D. Deb (2009), *Beyond Developmentality: Constructing Inclusive Freedom and Sustainability*, Abingdon, UK: Routledge, p. 87.

20. Set to equal the low end of the tax rates of the fifteen countries that are currently taxing carbon, to bring all countries up to this minimum level.

21. Set to be consistent with the proposed tax rate per barrel of oil. Based on the EIA carbon price scenario for 2030—$52 USD per metric tons of CO_2, and IEA forecasts for oil and natural gas prices for 2030—$97/bbl and $5.7/MMBtu

22. "Environmental Issues and UK Annual Reporting: Sustainable Business Initiative Turning Questions into Answers." ICAEW, 2015.

FOURTEEN

We Did It!

USA, 2050

Twenty years ago, I was giving a presentation on the rate at which renewable power would supplant conventional fossil fuel–driven power plants. After the presentation, a woman raised her hand and asked a long-winded question, in which she questioned my forecasts. She cited the high capital investment required to get some of these renewable energies off the ground, and the economies of scale necessary to get some of the new fuel sources to reach a critical level of adoption. She also questioned the technological readiness to make the changes occur. While she was absolutely correct that these factors would influence the rate of change, she generally underestimated the ability of corporations, governments, and consumers to adapt to a new reality.

Looking back over more than thirty years of management consulting, I can observe that I have generally tended to underestimate the rate of change in markets. It is always hard to predict change, and predicting a "hockey stick" change where a flat line suddenly veers upward at some point in the future, is a red flag to any consultant. Clients who want to grow a business or dig theirs out of a hole often make such projections and there is good reason to be skeptical about them. However, trends that are already underway tend to accelerate after they reach a "tipping point."

Through about 2020, the world was like the "boiling frog." Air pollution, water pollution, ozone depletion, and other externalities of the first era of globalization increased slowly, and it was hard for many people to see that their quality of life was deteriorating, and more especially their children faced a lower quality of life due to things we could control. We reached a tipping point when extreme weather had severe financial consequences and island populations needed to emigrate and become refugees in the EU and China.

Looking back, the fact is that industry *did* adapt. Manufacturing industries rose to the challenge and achieved aggressive targets with their top management's support, and the transformation came at an acceptable cost. The energy sector shifted to renewables. Process industries developed nontoxic substitutes. Discrete manufacturing industries enhanced measurement and reporting. Transportation industries increased R&D to compete in an era of intelligence and networked transportation. Retail industries embraced green building and recycling, and governments instituted balanced scorecards. Every actor played a role in achieving a sea change of thought and behavior. Society transitioned to a new social-industrial compact.

THE COST WAS REASONABLE

Under the regulatory regimes that were implemented for climate change and environmental balance, the cost of mitigation and compliance was affordable—below 2.5 percent of sales[1] (assuming sales is to a company what Gross Domestic Product [GDP] is to a country). An investment of 1 percent of GDP annually averted the worst effects.[2] Most countries and companies did not exceed an incremental expenditure of 2.5 percent.[3]

In addition, companies usually passed through some, if not all, of the costs of compliance in the form of higher prices, which halved the cost. The price impact of a $20/tonne carbon tax on twenty-seven sectors was generally a 1–2 percent increase in general wholesale prices (but a 20–30 percent increase in prices of gas and electricity).[4] Most economies still grew by an average of 2.4 percent per year. One country, Sweden, achieved positive growth while reducing its carbon footprint to zero.

The Standards Already Existed

Improving working conditions cost very little since most of the standards already existed, and implementing them took more *time* than *expense*.

To better understand how companies adapted to rising health, safety and environmental (HSE) legislation, I consult a specialist in this area. Anna manages the HSE department of a Canadian manufacturing company. She develops customized health and safety presentations and training programs, provides health and safety orientations for new employees, conducts safety training and education programs, and investigates accidents. She studied Industrial Technology in Venezuela before the Venezuelan government under Hugo Chavez and later Nicolás Maduro drove the economy to pieces, and she was forced to flee to Canada for economic and physical security.

Anna enumerates a wide range of corporate social responsibility standards, and suggests that all companies should be doing so voluntarily. In the area of corporate social responsibility (CSR), companies developed a corporate social responsibility policy, committed to the UN global compact, committed to the global reporting initiative, began maintaining a CSR incident log and actions as well as social targets and accountabilities, tolerated no drugs or alcohol policy, and bound subcontractors to globally accepted HSE standards.

Companies were able to meet the stricter requirements of workers' safety and working conditions by developing an HSE policy if they didn't already have one, developing and maintaining health and safety targets and accountabilities, developing a health and safety management system, committing to OHSAS 18001, documenting health and safety incidents and actions, developing a health and safety training policy, developing and maintaining a health and safety policy on working language, ensuring adequate worker protection equipment, and auditing health and safety audit practices.

In the area of Environmental Protection Systems, they developed environmental policies, committed to ISO 14001 and the Carbon Disclosure Project, began maintaining environmental targets and accountabilities, audited compliance to environmental policies, documented environmental incidents and actions, conducted environmental training and awareness, and maintained an environmental management system.

Making a leap in progress was not as hard as many had anticipated due to the availability of a wide range of existing standards that had been developed and promulgated by organizations such as DNV, Bureau Veritas, ISO, and NORSOK. Large numbers of companies were able to adopt these standards without major investment. For example, they implemented ISO 14000 (an environmental standard), ISO 14001 (a framework for setting environmental standards), OHSAS 18001 (an occupational health and safety management systems standard), ISO 31000 (a risk management framework), and OHAS (Occupational Health Advisory Service) 18000, which is specific to health and safety, and similar to ISO 14000.

Other standards were relatively easy to access and implement, for example: ISO 15544 (requirements and guidelines for emergency response), ISO 17776 (guidelines on tools and techniques for identification and assessment of hazards), ISO/IEC Guide 51:1999 (safety aspects to be included in various standards), ISO/IEC Guide 73:2002 (risk management vocabulary), and NORSOK Z-013 (Guidelines for Risk and Emergency Preparedness Analysis).[5]

ANCHOR PLAYERS DROVE CHANGE

"Anchor players" that defined the market due to their size and bargaining power took action. At the regional level, the EU mandated regulations and other regions and countries adopted the same regulations in order to be able to trade with the bloc. The United States adopted federal standards that coincided with those of its trading partners, particularly Canada and Mexico, in order to facilitate trade relations. And at the corporate level, giants in each industry took the lead in mandating change among their suppliers, companies such as Saudi Aramco in oil and gas, BASF in chemicals, Duke Energy in power, UPS in transportation, and Walmart in retail.

To get a clearer understanding of how these anchor players managed their extended supply chains to comply with the mandates and achieve the ecological, environmental, and social targets, I ask an expert on the extended supply chain how they did it. John, a thirty-year veteran of consulting and industry, is an expert in strategic sourcing, strategic alliance development, outsourcing, and business transformation. Most of his work has to do with driving change through the extended supply chain in industries such as electronics, telecommunications, consumer products, and oil, gas, and petrochemicals. He and I do some retrospective thinking on how companies adapted to increased ecological, environmental, and social expectations.

We conclude that many of the negative externalities that resulted in high D-Factors originated in the extended supply chain, and improvements in supply chain management were able to mitigate them. Efficient supply chain management techniques and practices reduced waste, and hence cost. A number of supply chain practices both reduced cost *and* improved the social and environmental footprint of the supply chain. Long-term supply agreements provided suppliers the conditions that justify investment in environmentally sound infrastructure. For example, Walmart found that longer-term agreements were effective and necessary to achieving its cost reduction and environmental goals. Closed-loop supply chains with "extended product responsibility" for reverse logistics facilitated reverse logistics flows that reduced waste and sometimes improved manufacturing efficiency by reusing existing materials. For example, Nike asked for used sneakers to be returned.[6] "Design for Disassembly" allows end-users to easily separate and return used component parts. Design for reverse logistics programs at HP, Kodak,[7] and other firms enabled efficient and reliable returns. Electronics firms pioneered much of the early work in reverse logistics due to the relatively high product price, high-volume logistics flows, stocking requirements, hazardous material content, and high rate of obsolescence of electronic products.

Nested packaging, wherein packaging from one tier becomes the product for the next, helped to transform packaging and shipping costs from indirect costs to direct costs. Using recycled packaging reduced the cost of packaging. Dell and Walmart use recycled packaging for laptops.[8] Even back in the 1920s, Henry Ford used incoming wood crates full of parts as floorboards of the car once the parts were unpacked.[9] Requiring suppliers to abide by sustainability principles and practices ensured consistency of practice across the extended supply chain. Most often this was accomplished by applying sustainability practices as criteria in the supplier selection process. In Alcatel's program, suppliers with a sustainability score of 60 percent or higher were considered "green suppliers." These suppliers, which represent 25 percent of all suppliers, on a base of $1.2b of external expenditure, were awarded preferential scoring on Requests for Proposal (RFPs).[10]

Involving suppliers early in the product design cycle ensured supplier participation in the early stages of program development and supply chain design, which enhanced their impact on future supply chain efficiency. Given that 70 percent of the cost becomes fixed in the first 20 percent of product and process design,[11] early supplier involvement was key to reducing long-run costs.

Minimizing supply risk by maintaining a portfolio of suppliers mitigated some negative consequences of many extended global supply chains. Even from the outset of the offshoring phenomenon, large manufacturers and distributors realized that their order and delivery lead times were so long that they needed to have backup sources closer to home. Many US retailers that were importing from China needed to place their orders in March for delivery in September through November in time to be placed in the stores for the Christmas season. Delivery cycles, which included a combination of ocean, transshipment and warehousing, rail, and truck modes across seven thousand miles limited the ability to change plans or return defective merchandise. Blyth, a marketer and manufacturer of decorative products, chose to keep at least one supplier closer to the point of sale in order to be able to respond more rapidly to trends in demand.[12] Many other Western manufacturers and retailers deliberately maintained a portfolio of suppliers including backup suppliers in relatively expensive countries to be capable of responding to changing customer desires and trends in a short time frame.[13] For companies in environmentally and socially conscious countries with stringent practices and laws, like Germany, maintaining local supply chains improved public perceptions and increased revenues.

Adopting a life-cycle approach identified links in the value chain that resulted in direct or indirect nonrecyclable waste. End-of-life disposal was the big opportunity in many cases. Improved processes for the maintenance and repair of consumables (such as toner cartridges in printers and oil and tires in cars) were prime opportunities for improving environmental and social footprint.

Here is how specific industries adapted.

OIL AND GAS COMPANIES REDEFINED THEIR MISSION

It would have been impossible to effect major change toward sustainability without the active participation of the oil, gas, and process industries. Oil and petroleum products, including petrochemicals, accounted for roughly 22 percent of total energy produced and approximately 10 percent of the GDP of major economies like the United States, Russia, and EU countries, and because they had everything to lose by migrating away from fossil fuels.[14] The industry's proactive engagement was an important milestone in reaching the tipping point. While other industries represented 1, 2, or 3 percent, oil and gas was literally a whole sector of the economy, so intensive collaboration between the public and private sector was needed to align around the goals.

For how the oil industry adapted to the new mandates, I consult Badr, my company's agent in Saudi Arabia. Badr explains to me that the oil industry's acceptance of, and active participation in the design of, a carbon tax broke an important deadlock in governing CO_2 emissions.

At first, the American Petroleum Institute (API) had been reluctant to agree to taxes or regulation that could negatively affect oil and gas companies' profit. At a Senate Finance Committee hearing on "Energy Tax Policy in 2016 and Beyond," API executive vice president, Louis Finkel argued for lower taxes, and, by extension, no carbon taxes, on American oil producers:

> America's oil and natural gas industry pays some of the highest tax rates among US businesses, generating billions of dollars every year in revenue for the federal government. Instead of calling for higher taxes that discourage domestic production, policymakers should follow pro-development energy policies that create jobs . . . and help our nation provide affordable and reliable energy for consumers. We need tax policies that encourage investment in America's abundant energy resources. It is because of our industry's investments that the US is the number one producer of oil and natural gas in the world while leading other countries in lowering carbon and other emissions. The industry also invests billions in low and zero emissions technologies to continue to drive down greenhouse gas emissions. This model has proven that we can protect the environment, grow our economy, and save consumers on average $1,300/year in energy costs.[15]

UK/Dutch and Canadian oil giants Shell, BP, and Suncor (Canada's largest oil company) were the first ones to accept the concept of a carbon tax: "Carbon pricing systems encourage the quickest and most efficient ways of reducing emissions widely," said Ben van Beurden, CEO of Royal Dutch Shell in 2015. "A global carbon price would help to unleash market forces and provide the right incentives for everyone to play their part. History has shown the power of market forces in making economies less energy intensive as people have found more efficient ways to use energy," explained Bob Dudley, CEO of BP, the same year. "Climate change is happening. We think a broad-based carbon price is the right answer," agreed Steve Williams, CEO of Suncor, a Canadian oil company.[16]

Badr talked about the transformation that occurred at ArabaCo. The company initially partially privatized its operations, then later largely divested oil and gas operations to refocus on renewable energy technologies. These changes brought more skills, talent, and growth to the Saudi economy than anyone ever imagined while the company was pursuing various "local content" and "Saudization" initiatives for decades prior to this.

ArabaCo broke the ice and was the bellwether that showed other oil companies that it was possible to find a financially rewarding future outside of oil and gas. The inevitability of clean energy had reached a tipping point of awareness that unleashed a flood of investments in renewable energy production, particularly in Europe and Asia. Natural gas now produces less than half the greenhouse gas (GHG) emissions of coal (450 vs. 950 CO_2eq/KwH),[17] and solar energy generates 9–20 percent of GHG emissions of natural gas. Wind energy generates no GHGs.

PROCESS INDUSTRIES DEVELOPED NONTOXIC ALTERNATIVES

The petrochemical industry had to adapt significantly, as it was a major cause of ecological risk, especially in "persistent" pollutants—the kind of stuff that never goes away, such as radioactive materials, chemicals that caused biological mutations, and plastic products that never biodegrade.

To get a better understanding of how the chemicals industry adapted, I consult my friend and technical expert Dr. Tom, who seemed to know everything about chemicals and toxicity. Dr. Tom spent forty years in drilling, completions, and production, and knew everything about how chemicals are used in those processes. He used to work at two of the major international oil and gas companies and three of the chemical companies that supply them. He holds patents, has written articles, basically he's a guru.

Dr. Tom reminded me that major new legislation had been enacted in response to two high-profile oil and gas industry incidents as early as 2010: the BP Deepwater Horizon accident, and a public outcry over possible health risks due to the fracturing fluids used in the exploration of shale gas and shale oil. In the United States, the latter resulted in an Environmental Protection Agency (EPA) subpoena that invoked the Toxic Substances Control Act (TSCA), the Clean Water Act, and the Resource Conservation and Recovery Act (RCRA) to force public disclosure of certain formulations that would have previously been considered "trade secrets" or "proprietary ingredients." Public oversight and regulation became tighter.

One of the most broad-reaching pieces of legislation following these events was the mandatory registration of chemical substances affected by the Registration, Evaluation, Authorisation of Chemicals (REACH) regulation (EC No. 1907/2006) and the related UN Globally Harmonised System (GHS).

Biocides were widely addressed in the legislation due to their wide scope of application—and chemical composition. Even lower-risk categories required registration, and if packaging and labeling was not correct, it could not be sold. Corrosion inhibitors and solvents are among the other products that are being significantly affected by REACH.

The European Chemicals Agency (ECHA) also published a list of Substances of Very High Concern (SVHCs). For example, 2-(2-methoxyethoxy)ethanol (DEGME), 2-(2-butoxyethoxy)ethanol (DEGBE), cyclohexane, methylenediisocyanate (MDI), and ammonium nitrate (used primarily in paints and coatings) were SVHCs. Starting in June 2011, any person or company that used more than one metric ton per year and the substance accounted for more than 0.1 percent of the mass of the object produced had to notify ECHA.

The classification, labelling and packaging of chemical substances (CLP) (European regulation (EC) No. 1272/2008), which implemented the Globally Harmonised System (GHS) in Europe, was related to REACH and had a similarly large impact. Its Article 5 classified products as toxic; very toxic; category 1 or 2 carcinogen; category 1 or 2 mutagen; toxic for reproduction (category 1 or 2); to be placed on the market for professional use or not to be used by the general public; and not to be authorized for marketing to the general public. CLP applied to all REACH registrants, including "downstream users" (nonmanufacturers).[18]

THE POWER SECTOR MIGRATED TO NATURAL GAS AND RENEWABLES

As power utilities consumed about 29 percent of total energy in the global economy, and were the largest single user of natural gas, their acceptance of the problem and willingness to implement the solutions was critical to progressing toward a sustainable paradigm.[19]

There was one person who knows everything I need to know about power generation. Dennis, an industry veteran and mechanical engineer who knows the guts of the power plants better than anybody, has helped me with projects worldwide for more than a decade.

In hindsight, Dennis said, getting the power companies to engage was not as hard as with the oil and gas companies, partly because natural gas became so cheap that it made economic sense for power companies to switch from coal to gas for their new generating plants. Also, there were many similarities in capital and operating expenditures and HSE considerations, which eased the transition.

Many cost and profit drivers between thermal power and renewables were similar. At a high level, capital requirements in both thermal power and renewables were driven by turbine generators, civil works, and real estate. Operating costs were driven by replacement parts and electrical storage and distribution. Ecological concerns and permitting affected the timing, size, and type of installations.

Conventional oil and gas operators adapted to some operational differences, especially in procurement, supply chain, technology evaluation, and financing. While conventional power procurement departments sourced primarily or even only from well-established suppliers, renewables suppliers paid closer attention to new suppliers, including those from emerging economies. While both thermal power technology and renewables providers focused on steam and gas turbines, renewables operators focused on "next generation" platforms and new models, and also closely track new component technologies that could affect the evolution of long-run unit cost. While thermal power projects were often government-funded, renewables providers used new models of collaboration such as operator-supplier partnerships and UN-sponsored carbon swaps. In the partnership model, intellectual property protection was more critical than in more mature conventional power generation supply chains. Real estate became a more important driver for wind, solar, and biomass-based technologies since extensive land or offshore space was required for solar or wind farms, or to grow the feedstock for conversion to biofuel.[20]

Technological advances narrowed the gap in costs for solar and wind generation compared to electricity generated from conventional power technologies. Also, federal, state and municipal governments incentives for clean (nonfossil-fuel–derived) power edged the industry over the tipping point.

DISCRETE MANUFACTURERS INCREASED THEIR ENERGY EFFICIENCY

How did other manufacturers make the change? As manufacturing industries used 14 percent of the total energy in the economy, it was important to get them onboard with the movement toward a sustainable future.[21] Automotive manufacturers in particular had a lot to lose by switching away from oil, so their engagement was critical. For a better understanding how manufacturers adapted, I talked with the person I knew who had both deep and long experience on the subject from multiple manufacturing industries and environments.

I chat with my friend Rich, a professor in Boston. Rich teaches and trains, and he has consistently held leadership positions in professional associations and taught, published and written, including serving on many editorial boards. He has too many recognitions, certifications, and awards to enumerate.

In retrospect, he says, the automotive industry experienced the most challenging adjustment to the new regulatory environment, and did the most lobbying to prevent it from happening or slow it down. Ultimately, Chinese manufacturers drove the unit cost down, especially for automotive batteries for electric and hybrid vehicles. Early entry and government contracts or other subsidies proved to be essential to reaching the low unit cost offered on the worldwide markets. Several governments made next-generation transportation a strategic technological priority, and that paid off for manufacturers in those countries.

Nonautomotive manufacturing companies used continuous improvement, supply chain management, and fuel substitution in their efforts to become more environmentally friendly and comply with environmental legislation.[22] Rich offered several examples.

Pfizer, the global pharmaceutical company, took a number of energy conservation decisions at its Freiburg (Germany) facility, ranging from the installation of solar panels to a geothermal heating and cooling system, heat recovery systems, adiabatic cooling, and automatic power shutdown systems in order to conserve energy. Pfizer ran an "Energy and Climate Change Programme" in which it reduced GHGs by 63 percent. A team developed a portfolio of about two hundred energy efficiency projects, which included replacing the old boiler with a new one that ran on biomass and wood pellets. The geothermal project had a payback of less than two years, reduced electricity consumption by 3,325 MwH and CO_2 emissions by 1,200 metric tons. The boiler replacement reduced CO_2 emissions by 5,500 tons per year and yielded credits that Pfizer sold in the European Emissions Trading System (ETS).[23]

Sandvik, a manufacturer of metal cutting tools, launched a cobalt recycling program that established a reverse logistics flow from customers backward to itself. The company organized procedures, packaging, and customer support. Not only did the program reduce waste at the customer, but it also reduced energy consumption in the manufacturing process for new tools, since recycled cobalt requires 75 percent less energy and generated 40 percent less CO_2 emissions.[24]

Improvements sometimes come easier, however, especially if improvement processes were embedded in companies' day-to-day operating paradigms. Alcoa made many incremental and local improvements through its conventional employee incentive programs. For example, Alcoa employees engage in environmental and social projects as part of their ongoing social and community service activities.

IKEA's voluntary supply chain greening relied on a team of supply chain professionals (including eighteen foresters whose job was to know the conditions in which all the wood that IKEA buys was grown) asked its suppliers detailed questions about their social and environmental practices and performance.[25]

Many of the negative externalities that resulted in high D-Factors originated in the extended supply chain, and improvements in supply chain management mitigated them. Efficient supply chain management techniques and practices reduce waste, and hence cost. For example, improving routing efficiency, delivering at off-peak hours, and migrating to newer more efficient engine technologies saved cost and reduce emissions. However, some cost reduction initiatives frequently created negative externalities, especially offshoring and outsourcing.

There were many opportunities to take advantage of supply chain practices that both reduced cost *and* improved social and environmental footprint of the supply chain. Long-term supply agreements provided suppliers the conditions that justify investment in environmentally sound infrastructure. For example, most manufacturers found that longer-term agreements were effective and necessary to achieving its cost reduction and environmental goals. Closed-loop supply chains with "extended product responsibility" for reverse logistics facilitated reverse logistics flows that reduced waste and sometimes improved manufacturing efficiency by reusing existing materials. For example, Nike asked for used sneakers to be returned.[26] "Design for Disassembly" allowed end-users to easily separate and return component parts with different recycling attributes. Design for Reverse Logistics programs at HP, Kodak,[27] and other firms enabled efficient and reliable returns. Electronics firms pioneered much of the early work in reverse logistics due to the relatively high product price, high-volume logistics flows, stocking requirements, hazardous material content, and high rate of obsolescence of electronic products.

Nested packaging, whereby packaging from one tier becomes the product for the next, helped to transform packaging and shipping costs from indirect costs to direct costs. Henry Ford used incoming wood crates full of parts as floorboards of the car once the parts were unpacked.[28] Using recycled packaging can reduce the cost of packaging. Dell and Walmart use recycled packaging for laptops.[29] Requiring suppliers to abide by sustainability principles and practices ensured consistency of practice across the extended supply chain. Most often this was accomplished by applying sustainability practices as weighted criteria in the supplier selection process. In Alcatel's program, suppliers with a score of 60 percent or higher were considered "green suppliers." These suppliers, which represent 25 percent of all suppliers, on a base of $1.2b of external expenditure, were awarded preferential scoring on RFPs.[30]

Involving suppliers early in the product design cycle ensured supplier participation in the early stages of program development and supply chain design, which enhanced their impact on future supply chain efficiency. Given that 70 percent of the cost becomes fixed in the first 20 percent of product and process design,[31] early supplier involvement was key to reducing long-run costs.

Minimizing supply risk by maintaining a portfolio of suppliers mitigated some negative consequences of many extended global supply chains. Even from the outset of the offshoring phenomenon, large manufacturers and distributors realized that their order and delivery lead times were so long that they needed to have backup sources closer to home. Many US retailers that were importing from China needed to place their orders in March for delivery in September through November in time to be placed in the stores for the Christmas season. Delivery cycles, which included a combination of ocean, transshipment and warehousing, rail, and truck modes across seven thousand miles limited the ability to change plans or return defective merchandise. Blyth, a marketer and manufacturer of decorative products, chose to keep at least one supplier closer to the point of sale in order to be able to respond more rapidly to trends in demand.[32] Many other Western manufacturers and retailers deliberately maintained a portfolio of suppliers including backup suppliers in relatively expensive countries to be capable of responding to changing customer desires and trends in a short time frame.[33] For companies in environmentally and socially conscious countries with stringent practices and laws, like Germany, maintaining local supply chains was good business that also tended to mitigate some of the destructive business behaviors that occur in parts of the "third world."

Adopting a life-cycle approach identified links in the value chain that could result in direct or indirect nonrecyclable waste. End-of-life disposal was the big opportunity in many cases. Maintenance, repair, and management of consumables (such as toner cartridges in printers and oil and tires in cars) were typical prime opportunities for improving environmental and social footprint.

I thank Rich for his insight and retrospective. These practices had in fact been very important to many companies as they made the transition to a more sustainable future.

TRANSPORTATION INDUSTRIES INCREASED R&D FOR INTELLIGENT NETWORKS

Transportation companies were key actors in the migration to better, more sustainable business practices. They accounted for 14 percent of the final consumption of all energy, mostly oil. When it comes to fleets and transport policy, I want the best advice so I contact my friend and colleague, Dan. Dan is an economist who assesses local, regional, state, and multistate economies in terms of economic impacts, competitiveness, target industries, strategic plans, and infrastructure investments. Dan analyzes the economics of transit and freight projects, including economic impact analysis and financial analysis, and helps develop sustainability and smart growth initiatives.

Through our chat it becomes clear to me that battery storage technology had the most impact on the evolution of vehicles. Cars basically became, in the words of Tony Seba, a professor at Stanford University, "computers on wheels," and battery technology gave them a two hundred plus mile range at $20,000 cost,[34] which allowed countries and cities to mandate a transition from fossil-fuel vehicles to electric vehicles to reduce air pollution. China targeted thirty-five million electric vehicles by 2025, and it worked.[35]

In hindsight, information technology rather than alternative fuels had the most effect on the environment in the transport industry. Uber, Lyft, and Didi created routing efficiencies and reduced the total number of vehicles on the road by 75 to 90 percent as more people decided they don't need a personal vehicle all the time.[36, 37, 38] Ridesharing applications reduced emissions, and each ride-sharing car removed 5.5 to 12.7 metric tons of GHG emissions annually.[39] The cumulative effect was a 10 percent reduction in CO_2 emissions from cars.[40]

Interconnected vehicles, which share information with each other wirelessly, and smart traffic control (intelligent transportation systems) further reduced the number of vehicles per person by increasing routing efficiency to the point where more individuals decided to forego a personal vehicle.

Finally, of course, the increasing use of drones for commercial delivery replaced some road vehicles and the pollutants that they generated.

RETAIL EMBRACED RECYCLING AND GREEN BUILDING

For anything involving green building, I rely on the opinion of my friend Fritz. Fritz is an energy efficiency expert with a lot of experience in the retail industry. He is a bigwig in an energy technology company that serves major retailers and commercial companies. The projects usually involve installation and management of multimeasure energy efficiency projects, including upgrading lighting, HVAC, refrigeration systems, compressed air, boiler, and steam systems. He is keen on technology, having been both an investor and a recipient of capital for clean energy projects.

We meet at O'Hare airport in Chicago. I point out that retailers made a huge impact quickly because of the "low-hanging fruit" in stores, restaurants, and office buildings. Waste was so obvious in many places, especially in emerging economies, that green improvements easily made by local staff generated immediate payback in many cases. "No-brainer" improvements that did not require investment or new equipment included: separating recyclables from trash, and recycling the recyclables (including food trash); reducing plastics consumption by, for example, offering individual utensils instead of giving every customer a complete packaged set of utensils; reducing plastics volume by, for example, cutting back on gross waste such as unnecessarily thick plastic bags and plastic bags inside plastic bags, and by using sporks instead of separate spoons and forks; and using bags that are recyclable in the locality.

Fritz doesn't disagree with my simplistic observations as a frequent international traveler, but is clearly disappointed in my laymen's view of what happened. He is more intrigued by the way in which companies adopted more technically sophisticated energy efficiency measures.

For new buildings, retailers were at liberty to adopt a plethora of new technology that helped them dramatically alter their energy usage profile. They adopted Leadership in Energy and Environmental Design (LEED) building standards, retrofitted lighting, and installed smart equipment and appliances.

Lighting was a prime target for energy efficiency. They used high efficiency, Consortium for Energy Efficiency (CEE)-specified, and European Commission (CE)-certified lamps and fixtures; retrofitted existing fixtures with high intensity fluorescent, induction lighting, and linear fluorescent lights; installed and retrofitted lighting for LED, which had three to ten times the life expectancy; phased out older incandescent and T12 fluorescent technology in favor of compact flourescent lamps (CFL) and T8 technology; took advantage of EU and US utility rebates and tax incentives; and implemented new energy-efficient technologies in networked and stand-alone relay panel systems, and daylighting controls.

They also looked to plumbing, glass, and even cement to improve energy efficiency. In plumbing, retailers took advantage of opportunities to upgrade existing water-consuming plumbing equipment with water-saving equipment, including using self-closing faucets rather than handle faucets or sensor faucets (self-closing faucets reduced water consumption by 30 percent); installed high-efficiency toilets that consumed 90 percent less water than the traditional ones; and used high-efficiency condensing boiler technology with stainless steel heat exchanger design to operate boilers at lower temperatures than typical hydronic boilers. In glass, retailers used e-glass, which cut energy losses in half compared to uncoated glass. They also used window film, which blocked heat from sun to reduce energy consumption almost in half, and retrofitted double-glazed windows directly on top of existing single-glazed windows, resulting in a triple glazed panel. Some also used glass, window, and film testing equipment to identify high-opportunity energy cost savings that would have otherwise gone unnoticed. In cement, some companies used newer varieties of Portland limestone cement, which decreased CO_2 emissions from manufacturing.

Retailers worldwide also adopted the EU's privacy laws, which were presaged by a 1995 Data Protection Initiative and supported by a 2012 Directive on the Protection of Personal Data. Privacy had previously been subordinated to growth and convenience. Location-based services via mobile phone had allowed vendors to track users' movements. Data enrichment, the aggregation and merging of public and private data sources on everything from credit history to real estate transactions and much more, had allowed companies to assemble highly detailed profiles of their customers and prospective customers, consisting of data that many of the customers did not want the vendor to know. The extraordinary power of internet tracking allowed companies to charge different prices to different consumers for the same or similar products, or even to charge the same consumer different prices at different moments, based on variables such as income, age, zip code, Internet service provider, or computer operating system. The EU directives put boundaries on these violations of personal data and allowed for "opt-in" instead of "get out!," and protected the privacy of consumers and employees.

NOTES

1. D. Leonard Downie et al. (2009), *Climate Change: A Reference Handbook*, ABC-CLIO, California, pp. 232–236.
2. *The Economics of Climate Change: The Stern Review*, Nicolas Stern, 2006.
3. Fischer and Morgenstern 2003, as cited in D. Leonard Downie et al. (2009), ABC-CLIO, California. See pp. 67–69 for details and pp. 232–236 for the references.
4. M. Common (1995), *Sustainability and Policy*, Cambridge University Press, p. 236.
5. David Jacoby, "Major Accident Prevention and Lessons Learned in Oil and Gas: Four Guidelines and Thirteen Best Practices for Governing Safe Supply Chains." Society of Petroleum Engineers. SPE-161035-PP.
6. M. Pullman, (2012), *Sustainability Delivered*, Business Expert Press.
7. M. Pullman, (2012), *Sustainability Delivered*, Business Expert Press.
8. Ibid.
9. Ibid.
10. T. Johnsen et al. (2014), *Purchasing and Supply Chain Management: A Sustainability Perspective*, Oxon, UK.
11. J. Jakubowski et al. (2014), "Design for Manufacturability in Virtual Environment: Using Knowledge Engineering," *Management and Production Engineering Review*, Volume 5, Number 1, pp. 3–10. DOI: 10.2478/mper-2014-0001.
12. David Jacoby (2005), "Courting the Consumer Creating Dynamic Brands in Retail and Consumer Goods." An Economist Intelligence Unit white paper sponsored by Information Resources, Inc., p. 11.
13. David Jacoby (2004), "State of Strategic Sourcing." Presentation to New England Supply Chain Conference and Exhibition, September 21.

14. Based on author's analysis of International Energy Agency data. For underlying data, see http://www.iea.org/sankey/.

15. Adapted from press release. F. Brindle (2016), "API Against Higher Tax on Oil and Gas," Editorial Assistant. https://www.energyglobal.com/downstream/refining/15062016/api-evp-louis-finkel-against-proposed-oil-natural-gas-tax-rise-3525/. Accessed June 15, 2017.

16. Tina Rosenberg (2017), "Guess Who's for a Carbon Tax Now." *New York Times.* April 11, 2017. https://nyti.ms/2olzx1R

17. IPCC (2014), *Climate Change 2014: Mitigation of Climate Change. Contribution of Working Group III to the Fifth Assessment Report of the Intergovernmental Panel on Climate Change,* Cambridge University Press, United Kingdom and New York, NY, USA, p. 539.

18. OTC 22005-PP. *Global Trade Restrictions and Related Compliance Issues Pertaining to Oil and Gas Production Chemicals.* David Jacoby, Boston Strategies International. Copyright 2011, Offshore Technology Conference. This paper was prepared for presentation at the Offshore Technology Conference held in Houston, Texas, USA, 2–5 May 2011.

19. Based on author's analysis of International Energy Agency data. For underlying data, see http://www.iea.org/sankey/.

20. David Jacoby (June 2012), "Road to Renewables: Bridging from Thermal Power to Renewable Power Supply Chains." GE News & Views.

21. Based on author's analysis of International Energy Agency data. For underlying data, see http://www.iea.org/sankey/.

22. S. Leon (2013), *Sustainability in SCM Casebook,* Pearson.

23. Ibid., p. 75.

24. S. Leon (2014), *Sustainability in Supply Chain Management Casebook: Applications in SCM,* Pearson Press.

25. D. Esty (2006), *Green to Gold,* Yale University Press.

26. M. Pullman (2012), *Sustainability Delivered,* Business Expert Press.

27. M. Pullman (2012), *Sustainability Delivered,* Business Expert Press.

28. Ibid.

29. Ibid.

30. T. Johnsen et al. (2014), *Purchasing and Supply Chain Management: A Sustainability Perspective,* Oxon, UK.

31. J. Jakubowski et al. (2014), "Design for Manufacturability In Virtual Environment: Using Knowledge Engineering," *Management and Production Engineering Review,* Volume 5, Number 1, pp. 3–10. DOI: 10.2478/mper-2014-0001.

32. David Jacoby (2005), "Courting the Consumer Creating Dynamic Brands in Retail and Consumer Goods." An Economist Intelligence Unit white paper sponsored by Information Resources, Inc., p. 11.

33. David Jacoby, "State of Strategic Sourcing." Presentation to New England Supply Chain Conference and Exhibition, September 21, 2004.

34. "All Fossil-Fuel Vehicles Will Vanish in 8 Years in Twin 'Death Spiral' for Big Oil and Big Auto." ImpactLab. May 18th, 2017. http://www.impactlab.net/2017/05/18/all-fossil-fuel-vehicles-will-vanish-in-8-years-in-twin-death-spiral-for-big-oil-and-big-auto/

35. *Reuters* (2017), "China Targets 35 Million Vehicle Sales by 2025, NEVs to Make up One-Fifth." April 25. https://www.reuters.com/article/us-china-autos-electric/china-targets-35-million-vehicle-sales-by-2025-nevs-to-make-up-one-fifth-idUSKBN17R086

36. *Time* (2016), "Self-Driving Cars Could Help Save the Environment—Or Ruin It. It Depends on Us", http://time.com/4476614/self-driving-cars-environment/. Accessed June 15, 2017.

37. *The Ecoguide* (2016), "How Green Is the Future of Self-Driving Cars?," http://theecoguide.org. http://whatsyourimpact.org/the-eco-guide/how-green-future-self-driving-cars. Accessed June 15, 2017.

38. Center for American Progress (2016), "The Impact of Vehicle Automation on Carbon Emissions Where Uncertainity Lies," https://www.Americanprogress.org/issues/green/reports/2016/11/18/292588/the-impact-of-vehicle-automation-on-carbon-emissions-where-uncertainty-lies/. Accessed June 15, 2017.

39. *Daily Caller* (2016), "Study: Uber and Lyft Reduce More CO2 Than Other Green Policies," http://dailycaller.com, https://phys.org/news/2016-07-car-mobility-decreases-greenhouse-gas.html and http://dailycaller.com/2016/07/20/study-uber-and-lyft-reduce-more-co2-than-other-green-policies/. Accessed June 15, 2017.

40. Motherboard (2016), "Uber Wants Us to Think It's Environmentally Friendly, But Is It?," https://motherboard.vice.com/en_us/article/is-uber-good-or-bad-for-the-environment. Accessed June 15, 2017.

Epilogue

2050

The transition to the new economy is over, and the dust has settled. Here is what has happened to the people you have met.

COLOMBIA

Camila Gabriela Moreno Sánchez, my company's consultant in Colombia, invented a new technology for energy storage and became rich by licensing the rights to Tesla. Juan Alejandro Muñoz Ramirez, my company's agent in Colombia, now works for Camila Gabriela. Ana Rosa, the coffee plantation owner in Colombia, manages a milling operation that her uncle started in Colombia. She employs 230 people and sells worldwide.

UNITED STATES

Alex, my analyst who crunched D-Factor numbers, helped me compile a worldwide database of D-Factors that are now used on product labels worldwide.

In the coffee business, Luuk Van den Berg, manager of the business of Koffie King, moved back to The Netherlands with his family and his blind dog to start a tea distribution company. The company now trades on the London Stock Exchange. Steve, the head of rail and intermodal operations in North America, was fired for accepting kickbacks from large shippers in exchange for large contracts. Nobody knows where he is now. Rob, the freight forwarder at UDS, bought a banana plantation in Costa Rica and happily retired there. And Patrick, the UDS driver, began day-trading and made millions of dollars last year, mostly on trades of UPS stock.

In the smartphone business, Bob Gannett, the freight forwarder for smartphones, appeared in a wildly popular smartphone television commercial. He is much more famous now than he ever was in freight forwarding. Jay, the senior vice president of MarkItDown, started a company that exported cheap American electronic products to China. The company folded after burning through $20 million of venture capital.

NIGERIA

In the Nigerian airline business, Lola, my consultant and analyst in Lagos, married a Nigerian pop superstar and has become his full-time agent. I went to her wedding and attend her husband's concert when I am in the area. Victoria Ademola, the airport ground operations supervisor in Lagos, leaked outlandish but true information about corruption among the air cargo operators to the Nigerian press. She is now in a witness protection program in Monaco. Omoteleola Ilesanmi, the manager of the airline ticket agency in Lagos, became a bank teller when most airline ticket sales had migrated to the Internet.

In the Nigerian oil trade, Dr. Fasasi Abdoullahi Agboola sold his engineering company to PetroFranc for $120 million and moved to Miami.

EGYPT

In the Middle Eastern airline business, Raj Balakrishnan, operations manager at Egypt Air, retired to great fanfare and started a second career as an emcee on *Egyptian Idol*. Kareem Al Latif, the airline operations manager in Cairo, joined a non-profit that combats oppression and extremism worldwide. His work is reputed to be instrumental in the establishment of a moderate regime in Syria and the rebuilding of Aleppo. Mahmoud, the freight forwarder in Cairo, switched careers. He now works in a retail store selling Samsung products.

In the Nigerian oil business, Ahmed Al Abadi, the CEO of the Egyptian national oil shipping company, stopped a pirate attack from his desktop using remote technologies. Miramax Pictures tapped him to reenact the event in a sequel of *Captain Phillips*. Mutasim Al Nuaimi, the coordinator of inbound shipping activities at the shipping arm of the Egyptian national oil company in Cairo, invested in a European technology company that specialized in shipboard gas sensors, which yielded a small fortune. Sameer Al Produsit, the production manager at the Egyptian national oil company, took an early retirement and received full pay for the next twenty years as part of his pension package.

SAUDI ARABIA

Badr Al Ebrahimi, my company's agent in Saudi Arabia, became an advisor to the King regarding Saudi Arabia's transformation from fossil fuels to renewables. Muammar Khan, the oil and gas trader in Saudi Arabia, now trades energy futures in the GCC interconnection grid. And Lou Lewis, the procurement manager at ArabaCo, returned to the UK, where he sits on multiple boards in a comfortable semiretirement. And Tom Brown, the specialist who accompanied me on the trip to ArabaCo, wrote multiple engineering textbooks about pumps, compressors, and bearings.

Darren O'Reilly and Sameer Al Al Produsit, the transportation manager and production managers at ArabaCo, were laid off when ArabaCo reorganized to become a financial holding company. After twenty-six and twenty-nine years of service, respectively, they enjoyed handsome retirement packages. Wassim Al Fakir entered politics and was famous until word of his fake credentials were exposed, at which point he was impeached. Hassan Al Nopayit was convicted of fraud in the settlement of a class action suit led by foreign contractors.

Akbar, the crazy driver in Saudi Arabia, got therapy and religion.

UNITED ARAB EMIRATES

George Kendrick, the general manager of Osiris, retired and moved to Kansas City and has volunteered for six-month missions with the UNDP in Cote d'Ivoire, Vietnam, and Sudan. Shamsul Shahwani, manager of Al Wasil Shipping in Abu Dhabi, returned to Pakistan and set up a freight agency there. The two procurement managers who plagiarized material were fired for indecent behavior at a company party.

FRANCE

François Valjean, importing agent at PetroFranc, spearheaded the wind-down of PetroFranc's oil and gas activities, then entered politics and became a cabinet member in two presidential administrations. Henri LeGrand de Genève, the Swiss banker who bankrolled the bulk petroleum imports coming into Marseille, went bankrupt while financing a transaction for a Zimbabwean drug dealer who reneged on his end of the deal.

Valentine Egbenedict, oil importer in France, became a naturalized French citizen and helped manage the conversion of petroleum-based filling stations into LPG and plug-in electric filling stations across France. Geoff de la Raffinerie, manager of the refinery near Marseille, became vice president of Downstream for ExtonOil's worldwide operations. André Marchand, the tanker company owner in France, sold his company and retired in the south of France. He trades in art and volunteers for Greenpeace.

Carmen Lefebvre, the customs broker, had two children and took a part-time job with an accounting firm.

GERMANY

Marcos Schmidt da Silva, the refinery manager in Germany, became an investment manager in a private equity firm. He focuses on early-stage energy technology investments. Ralf, the tanker fleet owner in Germany, is a senior executive at a German bulk shipping line. Marco, the customs agent in the port of Hamburg, continued being a customs agent in the port of Hamburg.

VIETNAM

Friedrich Werner, the coffee grower in Vietnam who introduced me to the bottled water distributor, developed a new variety of coffee bean with a unique taste that has become a global sensation. He is known around the world for his coffee advertisements.

Chi Pham, the investment banker in Vietnam, became independently wealthy, primarily on a bet he made on a startup company in Vietnam's telecommunications sector, and continued being an angel investor in hundreds of startup ventures in Vietnam. Nguyên Tràn, the water bottler in Ho Chi Minh City, got an MBA in tourism and operates a tour company in the Mekong Delta.

Arthur Waterman, executive at I-Shop, was laid off as part of the company's divestment of its Asian assets following a massive accounting scandal. He took a job with a Japanese retail firm that was expanding into Singapore. Tony Liu, the Chinese transport specialist with a branch in Vietnam, still manages freight shipments in and out of Vietnam and Southeast Asia, and is hoping for his technology investments to pay off someday.

MALAYSIA

Phang Hoàng, the water bottler in Malaysia, joined the staff of the International Management Institute in Singapore, where she coordinated conferences that helped transfer international best management practices to Southeast Asia. J.J., the bottled water distributor in Malaysia, married Phang Hoàng and became a personal trainer. Pei Chyi Ling, the conference marketer in Malaysia, joined her family's retail business and has helped expand operations internationally. Yoong Nim Chee, the trucking company operator in Kuala Lumpur, still runs the trucking company, and was nominated to an advisory committee for urban planning there, where he has become active in local politics.

CHINA AND TAIWAN

In smartphone production, Tony Tan, manager of the smartphone assembly operation in Shanghai, returned to the United States with his Chinese wife and two children, and became a founding partner in a company that developed controls for Honda's self-driving vehicles. Bill Ryan, the production manager at Micro-Electronica in Shanghai, accepted a job at LG as global product manager for all its electronics products.

In the smartphone logistics chain, Dan Middleman, the third-party logistics king who managed the flow of smartphones from Asia to the Americas, and was stationed in Shanghai, became the President of King Orient Lines in Los Angeles. Lemon Li Fung, who took care of warehousing and logistics for smartphones in Shanghai, was promoted to vice president in charge of marketing. Patrick Fang, the logistics coordinator in Shanghai, became a professor at the University of Tianjin, like his father. Qiandy Mao, the customs broker for smartphones in Shanghai, now works in the logistics department of Apple in Cupertino, California.

Tina Wang, my company's agent in China, appeared on the cover of the Chinese equivalent of *Forbes* magazine after placing more than two thousand Chinese engineers on projects in the Middle East, then became a government deputy minister responsible for export promotion.

Binbin, Tina Wang's assistant, went with Tina to her new job. Garry, the oil tycoon who Tina Wang introduced me to, invested heavily in Chinese wells that used fracking to increase flow rates. He became rich from the oil that the wells produced, but he couldn't find any buyers for his share of the wells after China voted to ban fracking.

Edgar and Wei, the windfarm developers in China, filed for fourteen patents related to wind turbine generators and started their own turbine gear company. Breeze is suing them in court for stealing the company's technology. Andy Wang and Mo and Yu, owners of RK Valve in Taiwan, have made RK Valve the largest player in the worldwide plug valve market.

Dick Zhào, my client at Canton Transport Corporation in Shenzhen, became an executive board member of CTC, where he worked even less than he had before. And Feng Xie, the wealthy shipping tycoon in the Beijing cigar club, doubled his fortune by investing in Chinese social media companies. He never once used social media himself.

As for me, I have already booked flights and hotels for the 2070 redux of this adventure!

Afterword

Based on the journey chronicled in this book, nine individually effective and collectively feasible taxes and incentives, mandates, and institutional recommendations can encourage a healthy and growing industrial and commercial global economy into the next century. The strategies are: 1) Carbon tax; 2) Clean power mandate; 3) Clean fuel mandate; 4) 100 percent recyclable consumer products mandate; 5) Mandatory reporting of public companies into the UN Global Compact; 6) Mandatory D-Factor labels on consumer goods; 7) Balanced scorecard of progressively tightening KPIs; 8) Human zoning principles, benchmarks, and standards; and 9) Reporting of the Genuine Progress Indicator.

The adoption, endorsement, and/or refinement of these recommendations by NGOs and associations would help to drive the change. Backing by organizations such as the United Nations—via the UN Framework Convention on Climate Change (UNFCCC), the UN Conference on Trade and Development (UNCTAD), the UN Development Program (UNDP), the Environment Programme (UNEP), and the UN Commission on Sustainable Development (UNCSD)—as well as the Carbon Disclosure Project, the Ellen MacArthur Foundation, the Global Environment Facility, the Global e-Sustainability Initiative, the Global Reporting Initiative, the International Energy Agency, the Organization for Economic Cooperation and Development, the US Agency for International Development, the World Health Organization (WHO), and the World Resources Institute, to name a few, would increase the chances of correcting the imbalances highlighted in this book. Regional, national, and local regulatory agencies, governmental bodies, and lenders should refine and adapt the recommendations to local circumstances, and set up pilot programs. Further work by authorities such as the US Department of Energy and Federal Energy Regulatory Commission, the Council of European Energy Regulators (an EU organ), and the Energy Regulators' Network (an ASEAN body), and lenders such as the African Development Bank, the Asian Development Bank, the International Finance Corporation (IFC), the EXIM Bank, the Inter-American Development Bank, and the World Bank, would ensure the proper consideration of local factors in the implementation of policies and standards. These organizations should also establish and implement, if they are a regulatory agency, appropriate and graduated sanctions for failure to meet the targets. Independent specialists, including consulting and accounting firms as well as technical experts and engineers, should measure, report on, and compare the KPIs presented in this book and on more specific and granular metrics, at the request of the regional, national, and local organizations listed previously, and with guidance from the concerned NGOs. The data should relate and compare actual performance to benchmarks and targets, and they need to make the data available publicly online.

The stakes are high and the effort is large, but human capacity and talent to adapt is larger. The roadmap is clear, and creativity, technology and innovation will fill in any gaps as they arise. The future will be clean and bright.

This book is a work of fiction. Any resemblance of the companies, products, or people in this book to real companies, products, bran ds, or people is purely coincidental. The coffee, airline ticket, bottled water, and smartphone (and any other) value chain descriptions in this book are fictional. Neither the conversion steps described, nor the conditions in specific countries, are guaranteed to be accurate, true to life, or representative of the countries in which they are depicted to take place.

That being said, the D-Factor figures have some basis in reality. The methodology and framework, as well as the calculations for 2010, are explained in my book *The High Cost of Low Prices: A Roadmap to Sustainable Prosperity* (Business Expert Press, 2017). The 2030 extrapolations are based on the scenarios presented in each chapter. And while this is not intended as a research paper, I have included some footnotes throughout the text so you may consult books, articles, and other citations pertaining to specific topics.

Some people have asked me to explain why I decided to write the book. By way of explanation, here is how I came to realize that these topics are important.

My first job after college was with a firm that helped shipping companies plan and execute global transportation services. At that time global trade was booming, and the more and farther we could transport anything, the better. Several large carriers had recently started "round-the-world" service in which carriers sent a fleet of ships entirely eastbound or entirely westbound. The concept and the name and the global ambition of it were all exciting, and there was money in both the shipping and the consulting services. I did not think about the fuel consumption, the carbon emissions, or the environmental impact. In that stage of globalization, air was generally clean wherever you went in the world, and although many countries were poor, few had deeply rooted problems of overdevelopment, pollution, or ecological damage.

After graduate school, I worked at another firm that essentially helped make mass manufacturing and distribution of cheap goods economical worldwide. We helped build global production and distribution networks that took advantage of the lowest cost for every component. We also helped transport companies vertically integrate and expand around the world. I made money, my clients made money, and consumers worldwide got access to lower-cost goods and blueberries in January. Everything was good. Some externalities of global supply chains were visible but seemed localized.

Later, during a major oil boom, I began helping oil, gas, power, and petrochemical companies increase output at lower unit costs. Most large oil companies during this period set up or invest in offshore rigs in developing countries and negotiated production agreements that harvested oil reserves in exchange for payments to the host country's government, with the result that those countries' consumers could earn enough money to buy the cheap goods coming off the round-the-world ships that my previous work had set up.

Through my cumulative work across the manufacturing, energy, chemicals, and power sectors, I began to experience with my own eyes, as I worked and traveled across the globe, cycles of production, consumption, and environmental and social impact, and the interrelationships between them.

The impact has become much more visible over the last few decades. Most of us are too busy to notice creeping change unless it grows to the point of causing serious problems such as health and safety hazards. And when we *do* notice it, many of us figure that we can't do anything about it anyway because we need to keep making a living and are powerless to do anything about it.

I hope this book to show that there is a path for addressing and resolving these issues, and we can follow that path while prospering, enjoying what we love most, and becoming even more connected with each other and ourselves.

Appendix

A BRIEF CHRONOLOGY OF EUROPEAN ENVIRONMENTAL REGULATION THROUGH 2017

Policies and Attitudes Toward Drilling, Fracking, Acidizing, and Leaching

Starting in 1998, Germany's Green Party advanced the push toward renewable energy. In 2001, Germany initiated a phaseout of nuclear energy by 2020. In 2007, the EU passed a resolution mandating 20 percent renewable energy by 2020. These major milestones have defined an aggressive, environmentally active culture throughout Europe. The United Kingdom and Denmark have also enacted progressive environmental legislation. The United Kingdom is aiming to reduce greenhouse gas emissions by 34 percent by 2020 nationally, without buying climate credits from abroad. In 2008 the United Kingdom adopted a Climate Change Act under which national reduction targets in terms of carbon budgets are adopted. Under the Climate Change Act the United Kingdom has adopted a conditional target to reduce emissions by around 50 percent by 2027 on the condition that the EU adopts common reduction targets post-2020. The Danish government's target is to reduce total Danish greenhouse gas emissions by 40 percent by 2020 compared with the 1990 level. The Danish government wants all sectors, including non-emissions trading system (ETS) sectors, to contribute with concrete and documented reductions up to 2020 and beyond. Its Energy Agreement of March 2012 includes such objectives as producing 35 percent of energy from renewable sources, a 12 percent reduction in energy consumption compared to 2006 and almost 50 percent of the Danish electricity consumption to be produced by wind power. All of Denmark's energy supply, including transport energy consumption, is to be based on renewable energy by 2050. As part of this, oil for heating purposes and coal are to be phased out by 2030, and electricity and heating supply is to be 100 percent covered by renewable energy by 2035.

Europe has taxed carbon usage for over twenty-five years. In 1992, Denmark was one of the world's first countries to introduce a CO_2 tax on both households' and businesses' energy consumption. With the aim of encouraging energy efficiency and switching toward fuels with less CO_2 content, the standard tax rate was set at €13.4 per ton CO_2. As of 2014, the tax rate was $31 per ton of CO_2. The Danish carbon tax covers all consumption of fossil fuels (natural gas, oil, and coal), with partial exemption and refund provisions for sectors covered by the EU ETS, energy-intensive processes, exported goods, fuels in refineries and many transport-related activities. Fuels used for electricity production are not taxed by the carbon tax, but instead a tax on electricity production applies.

The UK Climate Change Program was established in 2000 to meet both the country's Kyoto Protocol commitment. A Climate Change Levy is charged on natural gas, electricity, petroleum, and hydrocarbon gas in a liquid state, coal, lignite and coke supplied for lighting, heating and power purposes to business customers in the industrial, commercial, agricultural and public service sectors. The Climate Change Levy isn't charged on road fuel and other oils that are already subject to excise duty. It's also not charged on certain other supplies of energy, including supplies to domestic customers, charities when put to a nonbusiness use, and business customers who consume very small quantities of energy. In addition, the EU forces airlines to pay CO_2 fees.[1]

The EU pioneered emissions trading. The EU's ETS, which was started in 2005, includes all large CO_2 emitters in the EU such as power plants, large industrial installations, and platforms for oil and gas production. For every tonne of CO_2 emitted, companies must purchase allowances within an overall allowances cap. The allowances can be traded on a free market.

The EU has issued a federal mandate for disclosure of chemicals used, as the United States has done at the state level. In the EU, the Sevesa II Directive (Directive 96/82/EC) regulates the industrial use of hazardous chemicals such as cyanide,[2] and the Groundwater Directive (Directive 80/68/EEC) bans the use of cyanide that could result in the quality of groundwater being polluted. In 2000, the Water Framework Directive (2000/60/EC) replaced the Groundwater Directive.[3]

An EU Commission Recommendation[4] requires member states to ensure companies who are engaged in fracking operations to publicly disclose the type of chemical substances and the volume of water being used.[5] Bulgaria, France, and the Netherlands[6] have banned fracking.[7, 8] Denmark,[9] Lithuania,[10] and the United Kingdom[11] have instituted permitting and approval procedures that require the companies to submit plans, including details on use of chemicals, and have monitoring and reporting procedures. Poland actually loosened its restrictions in 2014, allowing drilling of wells under five thousand feet without environmental impact assessments.[12] The EC took Poland to the Court of Justice over the loosened standard, arguing that the Polish law does not comply with the Environmental Impact Assessment Directive (Directive 2011/92/EU), which should be used to determine whether projects require an assessment of waste products, effects on water and soil, use of natural resources, and the risk of accidents.[13]

The United Kingdom is concerned and potentially poised to take action. Fracking had been used in a single well in the United Kingdom until 2011, when seismic activity, which had never been recorded there before, was detected in the area. The government temporarily suspended all fracking until it completed an assessment of the seismic activity and determined how to mitigate such events in the future. Regulations now require review of seismic risk and faults in the area before a license to fracture wells can be issued.[14] As in the United States, fracking opponents seek to use local land-use laws to slow fracking, pointing to the need to secure landowners' consent before drilling horizontally through subsurface shale.[15]

Incentives for Air and Water Pollution, Recycling, Nature, and Intelligent Transport

In 2000, the European Parliament and the council introduced Directive 2006/21/EC after the Baia Mare cyanide spill, which covers the management of waste from extractive industries.[16] This directive and in particular Article 13(6) requires that "the concentration of weak acid dissociable cyanide in the pond is reduced to the lowest possible level using best available techniques." After May 2008, mines are prohibited from releasing wastewater that exceeds 10 ppm of weak acid dissociable cyanide. Article 14 also requires that mines establish financial guarantees to ensure cleanup after the mine is complete. European industry has come up with a voluntary "Cyanide Code"[17] that aims to reduce environmental impacts with third-party audits of a company's cyanide management.

Europe's Euro 6 vehicle emission standard, which came into effect in 2016, puts it on par with North America's Tier 2 standard, which has been in place since before 2011.[18, 19]

The EU leads the world in water pollution control by identifying and quantifying at the federal level specific standards for suspended solids (35 mg/l, 35 for more than 10,000 population equivalent [p.e.], and 60 for 2,000-100,000 p.e.), biochemical oxygen demand (25 mg/l. minimum percentage of reduction: 70-90), and chemical oxygen demand (25 mg/l minimum percentage of reduction: 75).[20]

In the EU, the framework for regulating water quality is set by directives. The Industrial Emissions Directive (IED) combines and restates seven previous Directives related to industrial emissions, including the Integrated Pollution Prevention and Control (IPPC) Directive.[21] These represent regional legislation that member nations are expected to transfer into national law.[22] They are binding to all member states but require a national process to be applied.[23]

The EU has mandated a 50 percent recycling target for household waste, to be enacted by 2020. This would represent a continuation of a trend from 23 percent in 2001 and 35 percent in 2010.[24] European countries recycle a lot more than countries from other regions except for South Korea. Germany recycles the most, at 65 percent. Other European countries recycle from 26 to 65 percent of their household waste. For reference, the United States recycles 35 percent and Mexico recycles only 5 percent of its household waste.

The EU passed a Timber Regulation in 2013 that regulates logging and deforestation. Consistent with strong EU enforcement mechanisms, the body negotiated with Romania to change its logging practices to comply with the Timber Regulation.[25]

Environmental law enforcement in Europe is stricter than the rest of the world, and the EU explicitly works to improve enforcement mechanisms. According to the World Rule of Law Index's list of most lawful countries, seven of the top ten countries are European.[26] The index measures constraints on government powers, the absence of corruption, open government, fundamental rights, order and security, regulatory and enforcement, civil justice, and criminal justice.

The EU is vocal and activist in matters of the environment. For example, key European Environment Networks hosted a conference in 2016 in Utrecht, the Netherlands comprised of 194 participants from thirty-five countries in Europe with the goal of strengthening compliance and enforcement of EU environment and nature laws.[27]

Eighty-seven percent of non-governmental organizations (NGOs) campaigning activity on ethical and sustainable consumption issues occurred in Europe, compared to 9 percent in North America and 4 percent in the rest of the world.[28] Within Europe, NGOs are most active in Northern Europe, especially Germany, the Netherlands, Sweden, Norway, and the United Kingdom.[29]

European citizens and communities are conscious and active in addressing societal environmental issues. For example, the Marine LitterWatch is a grassroots effort aimed at informing and empowering citizen networks throughout Europe to achieve the EU's ambitious Marine Strategy Framework's rating of "Good Environmental Status" of Europe´s seas by 2020.[30] Also, the European Innovation Partnership on Smart Cities and Communities' outreach program is geared toward increasing citizen engagement, especially through its Citizen Focus Action Cluster. In 2015 the Action Cluster presented a Toolkit on Principles and Enablers for Citizen Engagement. Subsequently it integrated many of those principles into a Manifesto on Citizen Engagement.[31]

Europe is also generally a leader in skills development, skills versus job matching, and sustainability of higher wages, according to Hays's Global Skills Index. Most European countries rank between 6 and 7 on the index, whereas the mean value globally was 5.0.[32] Even relatively smaller, mostly Northern, European cities such as Copenhagen (Denmark), Stockholm (Sweden), Basel (Switzerland), and Hanover (Germany) have carved out core competences that allow these communities to flourish in the global knowledge economy rather than decay like many rust belt communities in the United States.

Europe has the shortest workdays in the world. France was the early leader, shortening its work week to thirty-five hours years ago. Sweden piloted a six-hour workday in 2015.[33]

Europe's propensity to regulate extends beyond the energy industry. For example, Europe has an active program in Smart Transport, through which it intends to deploy a variety of technologies such as sensors, short- and long-distance transmitters and in-car equipment to alert drivers of roadway obstacles on a corridor between Rotterdam and Vienna via Frankfurt and Munich.[34] The Cooperative ITS Corridor was launched in 2013. Europe has also introduced technology to limit the speed at which trucks can travel and "intelligent speed adaptation" that advises vehicles of safe speed limits and has the capability to limit driving speed. The European Committee for Standardization (Comité Européen de Normalisation—CEN) has a technical committee (TC278—Road Traffic and Transport Telematics) focused on ITS standards issues. TC278 works in close cooperation with the ITS Technical Committee (TC204) of ISO. Many ITS standards items are developed in parallel by ISO/TC204 and CEN/TC278.

NOTES

1. Worldwatch Institute (2013), *State of the World 2013: Is Sustainability Still Possible?*, Island Press, Washington, DC.
2. Council Directive 96/82/EC of 9 December 1996 on the control of major-accident hazards involving dangerous substances.
3. Directive 2000/60/EC of the European Parliament and of the Council of 23 October 2000 establishing a framework for community action in the field of water policy (the Water Framework Directive).
4. EU Commission Recommendation of 22 January 2014 on minimum principles for the exploration and production of hydrocarbons (such as shale gas) using high-volume hydraulic fracturing (2014/70/EU).
5. Ibid.
6. The Netherlands' law is a temporary moratorium that will expire in 2020.
7. G. Erbach (2014), *Shale Gas and EU Energy Security Briefing*, p. 4. European Union. Accessed on August 9, 2018 at http://www.europarl.europa.eu/RegData/etudes/BRIE/2014/542167/EPRS_BRI(2014)542167_REV1_EN.pdf.
8. Shale Gas International (2015), "Dutch Government Bans Shale Till 2020," Shale Gas International, www.shalegas.international/2015/07/13/dutch-government-bans-shale-till-2020/.
9. Ibid., p. 4.
10. ChemicalWatch (2015), "Few EU Countries Clear on Disclosure of Fracking Chemicals: Poland Alone Has Current Drilling Operations," https://chemicalwatch.com/23097/few-eu-countries-clear-on-disclosure-of-fracking-chemicals
11. D. Carrington (2015), "Companies Will Be Legally Required to Reveal Chemicals Used for Fracking," *The Guardian*, www.theguardian.com/environment/2015/jan/14/companies-will-be-legally-required-to-reveal-chemicals-used-for-fracking
12. G. Erbach (2014), *Shale Gas and EU Energy Security, European Parliament*, p. 5

13. European Commission (2016), "Environmental Impact Assessment: Commission Refers POLAND to the Court of Justice of the EU over Inadequate Assessment of Exploratory Mining Drillings," http://europa.eu/rapid/press-release_IP-16-1454_en.htm

14. Countries Approach Fracking with Interest and Caution. (2014, January 2). *New York Law Journal*, http://www.clm.com/publication.cfm?ID=475#_ednref7. Accessed June 14, 2017.

15. Bloomberg (2013), "Fracking Opponents Find Lawyers Beat Superglue in Slowing Shale," www.bloomberg.com/news/articles/2013-10-16/fracking-opponents-find-lawyers-beat-superglue-in-slowing-shale

16. Directive 2006/21/EC of the European Parliament and of the Council of 15 March 2006 on the management of waste from extractive industries.

17. ICMI International Cyanide Management Code for the Manufacture, Transport and Use of Cyanide in the Production of Gold, www.cyanidecode.org.

18. Improving the conversions between the various passenger vehicle fuel economy/CO_2 emission standards around the world. The International Council on Clean Transportation. Retrieved June 2016.

19. TransportPolicy.Net

20. Municipal Effluent Disposal Standards. Point Sources of Pollution: Local Effects and Its Control.

21. Reporting Guidance for Business on Environmental Key Performance Indicators: a Consultation on Guidance for UK Businesses, July 2012. DEFRA.

22. European Commission (2015a).

23. Compendium of Water Quality Regulatory Frameworks: Which Water for Which Use?

24. http://www.eea.europa.eu/media/newsreleases/highest-recycling-rates-in-austria. Accessed June 14, 2017.

25. http://www.balkaninsight.com/en/article/romania-to-toughen-law-to-protect-forests. Accessed June 14, 2017.

26. http://worldjusticeproject.org/sites/default/files/roli_2015_0.pdf. Accessed June 14, 2017.

27. European Union Network for the Implementation and Enforcement of Environmental Law (2015), "IMPEL Study Confirms That Significant Challenges Remain in Implementing EU Environmental Law," www.impel.eu/impel-study-confirms-that-significant-challenges-remain-in-implementing-eu-environmental-law/. Accessed June 14, 2017.

28. US Sustainability (2016), "NGOs, Sustainability and the Rise of Ethical Consumerism," http://thesustainabilityalliance.us/ngos-sustainability-and-the-rise-of-ethical-consumerism/. Accessed June 14, 2017.

29. Ibid.

30. European Environment Agency (2015), "European Citizens to Help Tackle Marine Litter," www.eea.europa.eu/themes/coast_sea/marine-litterwatch/engaging-european-citizens. Accessed June 14, 2017.

31. EU Smartcities, "Manifesto on Citizen Engagement," https://eu-smartcities.eu/content/manifesto-citizen-engagement-online-consultation. Accessed June 14, 2017.

32. Hays (2015), "Labour Markets in a World of Continuous Change," The Hays Global Skills Index 2015, p. 7.

33. Science Alert (2015), "Sweden Is Shifting to a 6-Hour Work Day," www.sciencealert.com/sweden-is-shifting-to-a-6-hour-workday. Accessed June 14, 2017.

34. Roads and Bridges (2015), "Traffic Management: World's First International Smart Road Nearing Launch in Europe," www.roadsbridges.com/traffic-management-world%E2%80%99s-first-international-smart-road-nearing-launch-europe. Accessed June 14, 2017.

Bibliography

Daly, Herman E. (1996). *Beyond Growth: The Economics of Sustainable Development* . Boston: Beacon Press.

Friedman, Thomas L. (2008). *Hot, Flat, and Crowded: Why We Need a Green Revolution—and How It Can Renew America*. New York: Farrar, Straus and Giroux.

———. (2016). *Thank You for Being Late: An Optimist's Guide to Thriving in the Age of Accelerations* (First edition.). New York: Farrar, Straus and Giroux.

———. (2005). *The World Is Flat: A Brief History of the Twenty-First Century*. New York: Farrar, Straus and Giroux.

———. (1999). *The Lexus and the Olive Tree*. New York: Farrar, Straus, Giroux.

Goldratt, Eliyahu M., and Jeff Cox. (2004).*The Goal: A Process of Ongoing Improvement*. Great Barrington, MA: North River Press.

Jacoby, David (2017). *The High Cost of Low Prices: A Roadmap to Sustainable Prosperity* Business Expert Press.

———. (2012). *Optimal Supply Chain Management in Oil, Gas, and Power Generation*. PennWell.

———. (2009). *The Economist Guide to Supply Chain Management*. *The Economist* and Bloomberg Press.

Ostrom, Elinor. (1990). *Governing the Commons: The Evolution of Institutions for Collective Action*. Cambridge [England] ; New York: Cambridge University Press.

Rivoli, Pietra. (2005). *The Travels of a T-Shirt in the Global Economy: An Economist Examines the Markets, Power and Politics of World Trade*. Hoboken, NJ: John Wiley & Sons.

Sachs, J. (2005). *The End of Poverty: Economic Possibilities for Our Time*. New York: Penguin Press.

———. (2015). *The Age of Sustainable Development*. New York, NY: Columbia University Press.

The Story of Stuff. (2007). Written and created by Annie Leonard. Director, Louis Fox. Producer, Erica Priggen. Berkeley, CA: Free Range Studios. http://www.storyofstuff.com/

Index

About the Author

David Jacoby, president of Boston Strategies International, is an economist, management consultant, and trade policy advisor. He has lived and worked in dozens of countries worldwide in the last thirty years.

He wrote "Guide to Supply Chain Management" (*The Economist* 2009); *Optimal Supply Chain Management in Oil, Gas, and Power Generation* (PennWell 2012); *The High Cost of Low Prices: A Roadmap to Sustainable Prosperity* (Business Expert Press 2018); and *Trump, Trade and the End of Globalization* (Praeger 2018). Formerly, he taught operations management at Boston University's graduate school of business, served as a contributing editor at the Economist Intelligence Unit, and consulted to the World Bank and hundreds of government and private sector clients.

He holds an MBA from the Wharton School, a master's in international business from the Lauder Institute, and a bachelor of science in finance and economics from the University of Pennsylvania.

www.ingramcontent.com/pod-product-compliance
Ingram Content Group UK Ltd.
Pitfield, Milton Keynes, MK11 3LW, UK
UKHW041628031026
15587UKWH00001B/145

9 781498 556743